Fundamentals and Applications of Chemical Process

Fundamentals and Applications of Chemical Process

Contributors :
Caroline O'Sullivan,
Peter O'Hare, *et al.*

AURIS REFERENCE LTD.
London, UK

Fundamentals and Applications of Chemical Process
Contributors : Caroline O'Sullivan *and* Peter O'Hare, *et al.*

Auris Reference Ltd., UK

www.aurisreference.com

United Kingdom

Copyright 2016

Printed in 2017 for Sale in the Indian Subcontinent

Notice

Fundamentals and Applications of Chemical Process

ISBN: 978-1-78154-993-3

British Library Cataloguing in Publication Data
A CIP record for this book is available from the British Library

Exclusively distributed by CBS Publishers & Distributors Pvt. Ltd.

Sales & Distribution Rights only for India, Pakistan, Bangladesh, Sri Lanka, Nepal and Bhutan. This book is not to be sold outside these territories.

PREFACE

As chemical processes have grown more complex, so have the safety systems required to prevent accidents. Chemical Process, offers students and practitioners a more fundamental understanding of safety and the application required to safely design and manage today's sophisticated processes.

This edition continues the definitive standard of the previous editions. The content has been extensively updated to today's techniques and procedures, and two new chapters have been added. A new chapter on chemical reactivity provides the information necessary to identify, characterize, control, and manage reactive chemical hazards. A new chapter on safety procedures and designs includes new content on safely management, and specific procedures including hot work permits, lock-tag-try, and vessel entry. For upper level undergraduate or graduate level Chemical or Mechanical Engineering courses in chemical process safety, as well as for chemical and mechanical engineers in the beginning of their careers who are interested in improving chemical process safety. It can also serve as a guide for anyone interested in learning about chemical process safety, including high school teachers, firemen, environmentalists, OSHA regulators, EPA regulators, and others.

The only book designed as a text for teaching chemical process safety, this book provides a comprehensive introduction to the essential technical fundamentals of chemical process safety. Its emphasis on fundamentals is intended to help both the student and the practicing scientist to understand the concepts and to apply them in an appropriate manner.

This page left intentionally blank.

CONTENTS

LIST OF CONTRIBUTORS

Caroline O'Sullivan

School of Pharmacy, Cavanagh Building, University College Cork, Cork, Ireland; E-Mails: katie.ryan@ucc.ie (K.B.R.); a.crean@ucc.ie (A.M.C.)
and
Department of Chemical and Process Engineering, Cork Institute of Technology, Bishopstown, Cork, Ireland

Peter O'Hare

The Nanotechnology and Integrated BioEngineering Centre, University of Ulster at Jordanstown, Newtownabbey, Co Antrim, BT37 OQB, Northern Ireland; E-Mail: p.ohare@ulster.ac.uk

Greg Byrne

School of Electrical, Electronic & Mechanical Engineering, University College Dublin, Belfield, Dublin 4, Ireland; E-Mail: gregory.byrne@ucd.ie

Liam O'Neill

Research & Development, EnBIO, Carrigtohill, Cork, Ireland; E-Mail: liam.oneill@enbiomaterials.com

Katie B. Ryan

School of Pharmacy, Cavanagh Building, University College Cork, Cork, Ireland; E-Mails: katie.ryan@ucc.ie (K.B.R.); a.crean@ucc.ie (A.M.C.)

Abina M. Crean

School of Pharmacy, Cavanagh Building, University College Cork, Cork, Ireland; E-Mails: katie.ryan@ucc.ie (K.B.R.); a.crean@ucc.ie (A.M.C.)

This page left intentionally blank.

Chapter 1

CHEMICAL PROCESS: AN INTRODUCTION

In 1987, Robert M. Solow, an economist at the Massachusetts Institute of Technology, received the Nobel Prize in economics for his work in determining the sources of economic growth. Professor Solow concluded that the bulk of an economy's growth is the result of technological advances. This is especially true in the chemical industry, which is entering an era of more complex processes: higher pressure, more reactive chemicals, and exotic chemistry. More complex processes require more complex safety technology. Many industrialists even believe that the development and application of safety technology is actually a constraint on the growth of the chemical industry. As chemical process technology becomes more complex, chemical engineers will need a more detailed and fundamental understanding of safety. Since 1950, significant technological advances have been made in chemical process safety.

Today, safety is equal in importance to production and has developed into a scientific discipline that includes many highly technical and complex theories and practices.

The word "safety" used to mean the older strategy of accident prevention through the use

of hard hats, safety shoes, and a variety of rules and regulations. The main emphasis was on worker safety. Much more recently, "safety" has been replaced by "loss prevention." This term includes hazard identification, technical evaluation, and the design of new engineering features to prevent loss. The subject of this text is loss prevention, but for convenience, the words "safety" and "loss prevention" will be used synonymously throughout.

Chemical plants contain a large variety of hazards. First, there are the usual mechanical hazards that cause worker injuries from tripping, falling, or moving equipment. Second, there are chemical hazards. These include fire and explosion hazards, reactivity hazards, and toxic hazards.

In a scientific sense, a chemical process is a method or means of somehow changing one or more chemicals or chemical compounds. Such a chemical process can occur by itself or be caused by an outside force, and involves a chemical reaction of some sort. In an "engineering" sense, a chemical process is a method intended to be used in manufacturing or on an industrial scale (see Industrial Process) to change the composition of chemical(s) or material(s), usually using technology similar or related to that used in chemical plants or the chemical industry.

Neither of these definitions is exact in the sense that one can always tell definitively what is a chemical process and what is not; they are practical definitions. There is also significant overlap in these two definition variations. Because of the inexactness of the definition, chemists and other scientists use the term "chemical process" only in a general sense or in the engineering sense. However, in the "process (engineering)" sense, the term "chemical process" is used extensively. The rest of the article will cover the engineering type of chemical process.

Although this type of chemical process may sometimes involve only one step, often multiple steps, referred to as unit operations, are involved. In a plant, each of the unit operations commonly occur in individual vessels or sections of the plant called units. Often, one or more chemical reactions are involved, but other ways of changing chemical (or material) composition may be used, such as mixing or sepration process. The process steps may be sequential in time or sequential in space along a stream of flowing or moving material; see chemical plant. For a given amount of a feed (input) material or product (output) material, an expected amount of material can be determined at key steps in the process from empirical data and material balance calculations. These amounts can be scaled up or down to suit the desired capacity or operation of a particular chemical plant built for such a process. More than one chemical plant may use the same chemical process, each plant perhaps at differently scaled capacities. Chemical Processes like Distillation and Crystallization, goes back to Alchemy in Alexandria.

Such chemical processes can be illustrated generally as block flow diagram or in more detail as process flow diagram. Block flow diagrams show the units as blocks and the streams flowing between them as connecting lines with arrowheads to show direction of flow.

In addition to chemical plants for producing chemicals, chemical processes with similar technology and equipment are also used in oil refining and other refineries, natural gas processing,polymer and pharmaceutical manufacturing, food processing,and water and wastewater treatment.

UNIT PROCESSING IN CHEMICAL PROCESS

Unit processing is the basic processing in chemical engineering. Together with unit operation it forms the main principle of the varied chemical industries. Each genre of unit processing follows the same chemical law much as each genre of unit operations follows the same physical law.

Chemical engineering unit processing consists of the following important processes:

- Oxidation
- Reduction
- Hydrogenation
- Dehydrogenation
- Hydrolysis
- Hydration
- Dehydration
- Halogenation
- Nitrification
- Sulfonation
- Ammoniation
- Alkaline Fusion
- Alkylation
- Dealkylation
- Esterification
- Polymerisation
- Polycondesation
- Catalysis

As chemical process technology becomes more complex, chemical engineers will need a more detailed and fundamental understanding of safety. The authors of this book set out the fundamentals of chemical process safety in this introduction.

In 1987, Robert M. Solow, an economist at the Massachusetts Institute of Technology, received the Nobel Prize in economics for his work in determining the sources of economic growth. Professor Solow concluded that the bulk of an economy's growth is the result of technological advances.

It is reasonable to conclude that the growth of an industry is also dependent on technological advances. This is especially true in the chemical industry, which is entering an era of more complex processes: higher pressure, more reactive chemicals, and exotic chemistry.

More complex processes require more complex safety technology. Many industrialists even believe that the development and application of safety technology is actually a constraint on the growth of the chemical industry.

As chemical process technology becomes more complex, chemical engineers will need a more detailed and fundamental understanding of safety. H. H. Fawcett said, "To know is to survive and to ignore fundamentals is to court disaster."[1] This book sets out the fundamentals of chemical process safety.

Since 1950, significant technological advances have been made in chemical process safety. Today, safety is equal in importance to production and has developed into a scientific discipline that includes many highly technical and complex theories and practices. Examples of the technology of safety include

- Hydrodynamic models representing two-phase flow through a vessel relief
- Dispersion models representing the spread of toxic vapor through a plant after a release, and
- Mathematical techniques to determine the various ways that processes can fail and the probability of failure

Recent advances in chemical plant safety emphasize the use of appropriate technological tools to provide information for making safety decisions with respect to plant design and operation.

The word "safety" used to mean the older strategy of accident prevention through the use of hard hats, safety shoes, and a variety of rules and regulations. The main emphasis was on worker safety. Much more recently, "safety" has been replaced by "loss prevention." This term includes hazard identification, technical evaluation, and the design of new engineering features to prevent loss. The subject of this text is loss prevention, but for convenience, the words "safety" and "loss prevention" will be used synonymously throughout.

Safety, hazard, and *risk* are frequently used terms in chemical process safety. Their definitions are

- Safety or loss prevention: the prevention of accidents through the use of appropriate technologies to identify the hazards of a chemical plant and eliminate them before an accident occurs.
- Hazard: a chemical or physical condition that has the potential to cause damage to people, property, or the environment.
- Risk: a measure of human injury, environmental damage, or economic loss in terms of both the incident likelihood and the magnitude of the loss or injury.

Chemical plants contain a large variety of hazards. First, there are the usual mechanical hazards that cause worker injuries from tripping, falling, or moving equipment. Second, there are chemical hazards. These include fire and explosion hazards, reactivity hazards, and toxic hazards.

As will be shown later, chemical plants are the safest of all manufacturing facilities. However, the potential always exists for an accident of catastrophic proportions. Despite substantial safety programs by the chemical industry, headlines of the type shown in Figure 1-1 continue to appear in the newspapers.

Figure 1-1 Headlines are indicative of the public's concern over chemical safety.

1-1 Safety Programs

A successful safety program requires several ingredients, as shown in Figure1-2. These ingredients are

- System
- Attitude
- Fundamentals
- Experience
- Time
- You

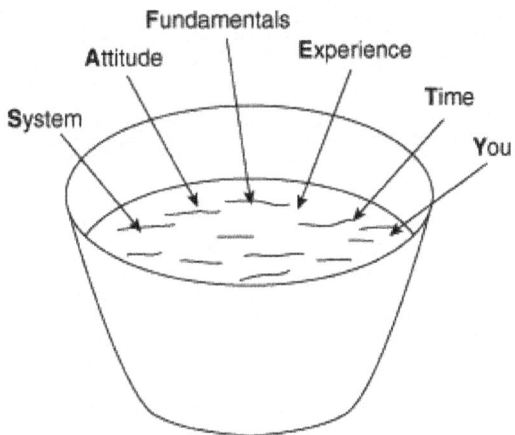

Figure 1-2 The ingredients of a successful safety program.

First, the program needs a system (1) to record what needs to be done to have an outstanding safety program, (2) to do what needs to be done, and (3) to record that the required tasks are done. Second, the participants must have a positive attitude. This includes the willingness to do some of the thankless work that is required for success. Third, the participants must understand and use the fundamentals of chemical process safety in the design, construction, and operation of their plants. Fourth, everyone must learn from the experience of history or be doomed to repeat it. It is especially recommended that employees (1) read and understand case histories of past accidents and (2) ask people in their own and other organizations for their experience and advice. Fifth, everyone should recognize that safety takes time. This includes time to study, time to do the work, time to record results (for history), time to share experiences, and time to train or be trained. Sixth, everyone (you) should take the responsibility to contribute to the safety program. A safety program must have the commitment from all levels within the organization. Safety must be given importance equal to production.

The most effective means of implementing a safety program is to make it everyone's responsibility in a chemical process plant. The older concept of identifying a few employees to be responsible for safety is inadequate by today's standards. All employees have the responsibility to be knowledgeable about safety and to practice safety.

It is important to recognize the distinction between a good and an outstanding safety program.

- A good safety program identifies and eliminates existing safety hazards.
- An outstanding safety program has management systems that prevent the existence of safety hazards.

A good safety program eliminates the existing hazards as they are identified, whereas an outstanding safety program prevents the existence of a hazard in the first place.

The commonly used management systems directed toward eliminating the existence of hazards include safety reviews, safety audits, hazard identification techniques, checklists, and proper application of technical knowledge.

1-2 Engineering Ethics

Most engineers are employed by private companies that provide wages and benefits for their services. The company earns profits for its shareholders, and engineers must provide a service to the company by maintaining and improving these profits. Engineers are responsible for minimizing losses and providing a safe and secure environment for the company's employees. Engineers have a responsibility to themselves, fellow workers, family, community, and the engineering profession. Part of this responsibility is described in the Engineering Ethics statement developed by the American Institute of Chemical Engineers (AICHE), shown in Table 1-1.

Table 1-1. American Institute of Chemical Engineers Code of Professional Ethics

Fundamental principles
Engineers shall uphold and advance the integrity, honor, and dignity of the engineering profession by 1. using their knowledge and skill for the enhancement of human welfare; 2. being honest and impartial and serving with fidelity the public, their employers, and clients; 3. striving to increase the competence and prestige of the engineering profession.
Fundamental canons
1. Engineers shall hold paramount the safety, health, and welfare of the public in the performance of their professional duties. 2. Engineers shall perform services only in areas of their competence. 3. Engineers shall issue public statements only in an objective and truthful manner. 4. Engineers shall act in professional matters for each employer or client as faithful agents or trustees, and shall avoid conflicts of interest. 5. Engineers shall build their professional reputations on the merits of their services. 6. Engineers shall act in such a manner as to uphold and enhance the honor, integrity, and dignity of the engineering profession. 7. Engineers shall continue their professional development throughout their careers and shall provide opportunities for the professional development of those engineers under their supervision.

Engineering ethics is the field of applied ethics and system of moral principles that apply to the practice of engineering. The field examines and sets the obligations by engineers to society, to their clients, and to the profession. As a scholarly discipline, it is closely related to subjects such as the philosophy of science, the philosophy of engineering, and the ethics of technology.

The 19th century and growing concern

The first Tay Bridge collapsed in 1879. At least sixty were killed.

As engineering rose as a distinct profession during the 19th century, engineers saw themselves as either independent professional practitioners or technical employees of large enterprises. There was considerable tension between the two sides as large industrial employers fought to maintain control of their employees.

In the United States growing professionalism gave rise to the development of four founding engineering societies: The American Society of Civil Engineers (ASCE) (1851), the American Institute of Electrical Engineers (AIEE) (1884), the American Society of Mechanical Engineers (ASME) (1880), and the American Institute of Mining Engineers (AIME) (1871).ASCE and AIEE were more closely identified with the engineer as learned professional, where ASME, to an extent, and AIME almost entirely, identified with the view that the engineer is a technical employee.

Even so, at that time ethics was viewed as a personal rather than a broad professional concern.

Turning of the 20th century and turning point

The Boston molasses disaster provided a strong impetus for the establishment of professional licensing and codes of ethics in the United States.

When the 19th century drew to a close and the 20th century began, there had been series of significant structural failures, including some spectacular bridge failures, notably the Ashtabula River Railroad Disaster(1876), Tay Bridge Disaster (1879), and the Quebec Bridg collapse (1907). These had a profound effect on engineers and forced the profession to confront shortcomings in technical and construction practice, as well as ethical standards.

One response was the development of formal codes of ethics by three of the four founding engineering societies. AIEE adopted theirs in 1912. ASCE and ASME did so in 1914. AIME did not adopt a code of ethics in its history.

Concerns for professional practice and protecting the public highlighted by these bridge failures, as well as the Boston molasses disaster (1919), provided impetus for another movement that had been underway for some time: to require formal credentials (Professional licensure in the US.) as a requirement to practice.

This involves meeting some combination of educational, experience, and testing requirements.

Over the following decades most American states and Canadian provinces either required engineers to be licensed, or passed special legislation reserving title rights to organization of professional engineers. The Canadian model is to require all persons working in fields of engineering that posed a risk to life, health, property, the public welfare and the environment to be licensed, and all provinces required licensing by the 1950s.

The US model has generally been only to require those practicing independently (*i.e.* consulting engineers) to be licensed, while engineers working in industry, education, and sometimes government need not be licensed. This has perpetuated the split between professional engineers and those in industry. Professional societies have adopted generally uniform codes of ethics. On the other hand technical societies have generally not adopted these, but instead sometimes offer ethics education and resources to members similar to those of the professional societies. This is not uniform, and the question of who is to be held in the highest regard: the public or the employer, is still an open one in industry, and sometimes in professional practice.

Recent Developments

William LeMessurier's response to design deficiencies uncovered after construction of the Citigroup Center is often cited as an example of ethical conduct.

Efforts to promote ethical practice continue. In addition to the professional societies and chartering organizations efforts with their members, the Canadian Iron Ring and American Order of the Engineer trace their roots to the 1907 Quebec Bridge collapse. Both require members to swear an oath to uphold ethical practice and wear a symbolic ring as a reminder.

In the United States, the National Society of Professional Engineers released in 1946 its Canons of Ethics for Engineers and Rules of Professional Conduct, which evolved to the current Code of Ethics, adopted in 1964. These requests ultimately led to the creation of the Board of Ethical Review in 1954. Ethics cases rarely have easy answers, but the BER's nearly 500 advisory opinions have helped bring clarity to the ethical issues engineers face daily.

Currently, bribery and political corruption is being addressed very directly by several professional societies and business groups around the world However, new issues have arisen, such as offshoring, sustainable development, and environmental protection, that the profession is having to consider and address.

General Principles

"Engineers, in the fulfillment of their professional duties, shall hold paramount the safety, health, and welfare of the public"

—National Society of Professional Engineers,

"A practitioner shall, regard the practitioner's duty to public welfare as paramount."

—Professional Engineers Ontario,

Codes of engineering ethics identify a specific precedence with respect to the engineer's consideration for the public, clients, employers, and the profession.

Many engineering professional societies have prepared codes of ethics. Some go back to the early decades of the twentieth century. These have been incorporated to a greater or lesser degree into the regulatory laws of several jurisdictions. While these statements of general principles served as a guide, engineers still require sound judgment to interpret how the code would apply to specific circumstances.

The general principles of the codes of ethics are largely similar across the various engineering societies and chartering authorities of the world, which further extend the code and publish specific guidance. The following is an example from the American Society of Civil Engineers:

1. Engineers shall hold paramount the safety, health and welfare of the public and shall strive to comply with the principles of sustainable development in the performance of their professional duties.

2. Engineers shall perform services only in areas of their competence.

3. Engineers shall issue public statements only in an objective and truthful manner.

4. Engineers shall act in professional matters for each employer or client as faithful agents or trustees, and shall avoid conflicts of interest.

5. Engineers shall build their professional reputation on the merit of their services and shall not compete unfairly with others.

6. Engineers shall act in such a manner as to uphold and enhance the honor, integrity, and dignity of the engineering profession and shall act with zero-tolerance for bribery, fraud, and corruption.

7. Engineers shall continue their professional development throughout their careers, and shall provide opportunities for the professional development of those engineers under their supervision.

Obligation to Society

The paramount value recognized by engineers is the safety and welfare of the public. As demonstrated by the following selected excerpts, this is the case for professional engineering organizations in nearly every jurisdiction and engineering discipline:

- Institute of Electrical and Electronics Engineers: "We, the members of the IEEE, ... do hereby commit ourselves to the highest ethical and professional conduct and agree: 1. to accept responsibility in making decisions consistent with the safety, health and welfare of the public, and to disclose promptly factors that might endanger the public or the environment;"

- Institution of Civil Engineers: "Members of the ICE should always be aware of their overriding responsibility to the public good. A member's obligations to the client can never override this, and members of the ICE should not enter undertakings which compromise this responsibility. The 'public good' encompasses care and respect for the environment, and for humanity's cultural, historical and archaeological heritage, as well as the primary responsibility members have to protect the health and well being of present and future generations."

- Professional Engineers Ontario: "A practitioner shall, regard the practitioner's duty to public welfare as paramount."

- National Society of Professional Engineers: "Engineers, in the fulfillment of their professional duties, shall: Hold paramount the safety, health, and welfare of the public."

- American Society of Mechanical Engineers: "Engineers shall hold paramount the safety, health and welfare of the public in the performance of their professional duties."

- Institute of Industrial Engineers: "Engineers uphold and advance the integrity, honor and dignity of the engineering profession by: 2. Being honest and impartial, and serving with fidelity the public, their employers and clients."

- American Institute of Chemical Engineers: "To achieve these goals, members shall hold paramount the safety, health and welfare of the public and protect the environment in performance of their professional duties."

- American Nuclear Society: "ANS members uphold and advance the integrity and honor of their professions by using their knowledge and skill for the enhancement of human welfare and the environment; being honest and impartial; serving with fidelity the public, their employers, and their clients; and striving to continuously improve the competence and prestige of their various professions."

RESPONSIBILITY OF ENGINEERS

The engineer recognizes that the greatest merit is the work and exercises his profession committed to serving society, attending to the welfare and progress of the majority. By transforming nature for the benefit of mankind, the engineer must increase his awareness of the world as the abode of man, his interest in the universe as a guarantee of overcoming his spirit, and knowledge of reality to make the world fairer and happier. The engineer should reject any paper that is intended to harm the general interest, thus avoiding a situation that might be hazardous or threatening to the environment, life, health, or other rights of human beings. It is an inescapable duty of the engineer to uphold the prestige of the profession, to ensure its proper discharge, and to maintain a professional demeanor rooted in ability, honesty, fortitude, temperance, magnanimity, modesty, honesty, and justice; with the consciousness of individual well-being subordinate to the social good. The engineer and his employer must ensure the continuous improvement of his knowledge, particularly of his profession, disseminate his knowledge, share his experience, provide opportunities for education and training of workers, provide recognition, moral and material support to the school where he studied, thus returning the benefits and opportunities he and his employer have received. It is the responsibility of the engineer to carry out his work efficiently and to support the law. In particular, he must ensure compliance with the standards of worker protection as provided by the law. As a professional, the engineer is expected to commit himself to high standards of conduct (NSPE).

Whistleblowing

The space shuttle challanger disastris used as a case study of whistleblowing and organizational behavior including group think.

Main Article: Whistleblower

A basic ethical dilemma is that an engineer has the duty to report to the appropriate authority a possible risk to others from a client or employer failing to follow the engineer's directions. According to first principles, this duty overrides the duty to a client and/or employer. An engineer may be disciplined, or have their license revoked, even if the failure to report such a danger does not result in the loss of life or health.

In many cases, this duty can be discharged by advising the client of the consequences in a forthright matter, and ensuring the client takes the engineer's advice. However, the engineer must ensure that the remedial steps are taken and, if they are not, the situation must be reported to the appropriate authority. In very rare cases, where even a governmental authority may not take appropriate action, the engineer can only discharge the duty by making the situation public. [] As a result, whistleblowing by professional engineers is not an unusual event, and courts have often sided with engineers in such cases, overruling duties to employers and confidentiality considerations that otherwise would have prevented the engineer from speaking out.

Conduct

There are several other ethical issues that engineers may face. Some have to do with technical practice, but many others have to do with broader considerations of business conduct. These include:

- Relationships with clients, consultants, competitors, and contractors
- Ensuring legal compliance by clients, client's contractors, and others
- Conflict of interest
- Bribery and kickbacks, which also may include:
- Gifts, meals, services, and entertainment
- Treatment of confidential or proprietary information
- Consideration of the employer's assets
- Outside employment/activities (Moonlighting)

Some engineering societies are addressing environmental protection as a stand-alone question of ethics.

The field of business ethics often overlaps and informs ethical decision making for engineers.

Case Studies and Key Individuals

Petroski notes that most engineering failures are much more involved than simple technical mis-calculations and involve the failure of the design process or management culture. However, not all engineering failures involve ethical issues. The infamous collapse of the first Tacoma Narrows Bridge, and the losses of the Mars Polar Lander and Mars Climate Orbiter were technical and design process failures.

These episodes of engineering failure include ethical as well as technical issues.

- Space Shuttle Columbia disaster (2003)
- Space Shuttle Challenger disaster (1986)
- Therac-25 accidents (1985 to 1987)
- Chernobyl disaster (1986)
- Bhopal disaster (1984)
- Kansas City Hyatt Regency walkway collapse (1981)
- Love Canal (1980), Lois Gibbs
- Three Mile Island accident(1979)
- Citigroup Center (1978),

- Ford Pinto safety problems (1970s)
- Minamata disease (1908–1973)
- Chevrolet Corvair safety problems (1960s), Ralph Nader, and Unsafe at Any Speed
- Boston molasses disaster (1919)
- Quebec Bridge collapse (1907), Theodore Cooper
- Johnstown Flood(1889), South Fork Fishing and Hunting Club
- Tay Bridge Disaster (1879), Thomas Bouch, William Henry Barlow, and William Yollan
- Ashtabula River Railroad Disaster (1876), Amasa Stone

CODE OF ETHICS OF ENGINEERS

Engineers Act

(chapter I-9, s. 7)

Professional Code

(chapter C-26, s. 87)

DIVISION I
GENERAL PROVISIONS

1.01. This Regulation is made pursuant to section 87 of the Professional Code (chapter C-26).

R.R.Q., 1981, c. I-9, r. 3, s. 1.01.

1.02. In this Regulation, unless the context indicates otherwise, the word "client" means a person to whom an engineer provides professional services, including an employer.

R.R.Q., 1981, c. I-9, r. 3, s. 1.02.

1.03. The Interpretation Act (chapter I-16) applies to this Regulation.

R.R.Q., 1981, c. I-9, r. 3, s. 1.03.

DIVISION II
DUTIES AND OBLIGATIONS TOWARDS THE PUBLIC

2.01. In all aspects of his work, the engineer must respect his obligations towards man and take into account the consequences of the performance of his work on the environment and on the life, health and property of every person.

R.R.Q., 1981, c. I-9, r. 3, s. 2.01.

2.02. The engineer must support every measure likely to improve the quality and availability of his professional services.

R.R.Q., 1981, c. I-9, r. 3, s. 2.02.

2.03. Whenever an engineer considers that certain works are a danger to public safety, he must notify the Ordre des ingénieurs du Québec or the persons responsible for such work.

R.R.Q., 1981, c. I-9, r. 3, s. 2.03.

2.04. The engineer shall express his opinion on matters dealing with engineering only if such opinion is based on sufficient knowledge and honest convictions.

R.R.Q., 1981, c. I-9, r. 3, s. 2.04.

2.05. The engineer must promote educational and information measures in the field in which he practises.

R.R.Q., 1981, c. I-9, r. 3, s. 2.05.

DIVISION III
DUTIES AND OBLIGATIONS TOWARDS CLIENTS

§1. *General provisions*

3.01.01. Before accepting a mandate, an engineer must bear in mind the extent of his proficiency and aptitudes and also the means at his disposal to carry out the mandate.

R.R.Q., 1981, c. I-9, r. 3, s. 3.01.01.

3.01.02. In cases where it is in his client's interest, the engineer shall retain the services of experts after having obtained his client's authorization, or he shall advise the latter to do so.

R.R.Q., 1981, c. I-9, r. 3, s. 3.01.02; O.C. 2566-84, s. 1.

3.01.03. An engineer must refrain from practising under conditions or in circumstances which could impair the quality of his services.

R.R.Q., 1981, c. I-9, r. 3, s. 3.01.03.

3.01.04. An engineer must at all times acknowledge his client's right to consult another engineer and, in such cases, he must offer his cooperation to the latter.

O.C. 2566-84, s. 2.

§2. *Integrity*

3.02.01. An engineer must fulfill his professional obligations with integrity.

R.R.Q., 1981, c. I-9, r. 3, s. 3.02.01.

3.02.02. An engineer must avoid any misrepresentation with respect to his level of competence or the efficiency of his own services and of those generally provided by the members of his profession.

R.R.Q., 1981, c. I-9, r. 3, s. 3.02.02.

3.02.03. An engineer must, as soon as possible, inform his client of the extent and the terms and conditions of the mandate entrusted to him by the latter and obtain his agreement in that respect.

R.R.Q., 1981, c. I-9, r. 3, s. 3.02.03.

3.02.04. An engineer must refrain from expressing or giving contradictory or incomplete opinions or advice, and from presenting or using plans, specifications and other documents which he knows to be ambiguous or which are not sufficiently explicit.

R.R.Q., 1981, c. I-9, r. 3, s. 3.02.04.

3.02.05. An engineer must inform his client as early as possible of any error that might cause the latter prejudice and which cannot be easily rectified, made by him in the carrying out of his mandate.

R.R.Q., 1981, c. I-9, r. 3, s. 3.02.05.

3.02.06. An engineer must take reasonable care of the property entrusted to his care by a client and he may not lend or use it for purposes other than those for which it has been entrusted to him.

R.R.Q., 1981, c. I-9, r. 3, s. 3.02.06.

3.02.07. Where an engineer is responsible for the technical quality of engineering work, and his opinion is ignored, the engineer must clearly indicate to his client, in writing, the consequences which may result therefrom.

R.R.Q., 1981, c. I-9, r. 3, s. 3.02.07.

3.02.08. An engineer shall not resort nor lend himself to nor tolerate dishonest or doubtful practices in the performance of his professional activities.

R.R.Q., 1981, c. I-9, r. 3, s. 3.02.08; O.C. 2566-84, s. 3.

3.02.09. An engineer shall not pay or undertake to pay, directly or indirectly, any benefit, rebate or commission in order to obtain a contract or upon the carrying out of engineering work.

R.R.Q., 1981, c. I-9, r. 3, s. 3.02.09.

3.02.10. An engineer must be impartial in his relations between his client and the contractors, suppliers and other persons doing business with his client.

R.R.Q., 1981, c. I-9, r. 3, s. 3.02.10.

§3. *Availability and diligence*

3.03.01. An engineer must show reasonable availability and diligence in the practice of his profession.

R.R.Q., 1981, c. I-9, r. 3, s. 3.03.01.

3.03.02. In addition to opinion and counsel, the engineer must furnish his client with any explanations necessary to the understanding and appreciation of the services he is providing him.

R.R.Q., 1981, c. I-9, r. 3, s. 3.03.02.

3.03.03. An engineer must give an accounting to his client when so requested by the latter.

R.R.Q., 1981, c. I-9, r. 3, s. 3.03.03.

3.03.04. An engineer may not cease to act for the account of a client unless he has just and reasonable grounds for so doing. The following shall, in particular, constitute just and reasonable grounds:

(a) the fact that the engineer is placed in a situation of conflict of interest or in a circumstance whereby his professional independence could be called in question;

(b) inducement by the client to illegal, unfair or fraudulent acts;

(c) the fact that the client ignores the engineer's advice.

R.R.Q., 1981, c. I-9, r. 3, s. 3.03.04.

3.03.05. Before ceasing to exercise his functions for the account of a client, the engineer must give advance notice of withdrawal within a reasonable time.

R.R.Q., 1981, c. I-9, r. 3, s. 3.03.05.

§4. *Seal and signature*

R.R.Q., 1981, c. I-9, r. 3, Div. III, Sd. 4; O.C. 2566-84, s. 4.

3.04.01. An engineer must affix his seal and signature on the original and the copies of every engineering plan and specification prepared by himself or prepared under his immediate control and supervision by persons who are not members of the Order.

An engineer may also affix his seal and signature on the original and the copies of documents mentioned in this section which have been prepared, signed and sealed by another engineer.

An engineer must not affix his seal and signature except in the cases provided for in this section.

R.R.Q., 1981, c. I-9, r. 3, s. 3.04.01; O.C. 2566-84, s. 4.

3.04.02. An engineer must affix his signature on the original and the copies of every written consultation and opinion, measurement, layout, report, computation, study, drawing and specification prepared by himself or prepared under his immediate control and supervision by persons who are not members of the Order.

An engineer may also affix his signature on the original and the copies of documents mentioned in this section which have been prepared and signed by another engineer.

O.C. 2566-84, s. 4.

§5. *Independence and impartiality*

3.05.01. An engineer must, in the practice of his profession, subordinate his personal interest to that of his client.

R.R.Q., 1981, c. I-9, r. 3, s. 3.05.01.

3.05.02. Any engineer must ignore any intervention by a third party which could influence the performance of his professional duties to the detriment of his client.

Without restricting the generality of the foregoing, an engineer shall not accept, directly or indirectly, any benefit or rebate in money or otherwise from a supplier of goods or services relative to engineering work which he performs for the account of a client.

R.R.Q., 1981, c. I-9, r. 3, s. 3.05.02.

3.05.03. An engineer must safeguard his professional independence at all times and avoid any situation which would put him in conflict of interest.

R.R.Q., 1981, c. I-9, r. 3, s. 3.05.03.

3.05.04. As soon as he ascertains that he is in a situation of conflict of interest, the engineer must notify his client thereof and ask his authorization to continue his mandate.

R.R.Q., 1981, c. I-9, r. 3, s. 3.05.04.

3.05.05. An engineer shall share his fees only with a colleague and to the extent where such sharing corresponds to a distribution of services and responsibilities.

R.R.Q., 1981, c. I-9, r. 3, s. 3.05.05.

3.05.06. In carrying out a mandate, the engineer shall generally act only for one of the parties concerned, namely, his client. However, where his professional duties require that he act otherwise, the engineer must notify his client thereof. He shall accept the payment of his fees only from his client or the latter's representative.

R.R.Q., 1981, c. I-9, r. 3, s. 3.05.06.

§6. Professional secrecy

3.06.01. An engineer must respect the secrecy of all confidential information obtained in the practice of his profession.

R.R.Q., 1981, c. I-9, r. 3, s. 3.06.01.

3.06.02. An engineer shall be released from professional secrecy only with the authorization of his client or whenever so ordered by law.

R.R.Q., 1981, c. I-9, r. 3, s. 3.06.02.

3.06.03. An engineer shall not make use of confidential information to the prejudice of a client or with a view to deriving, directly or indirectly, an advantage for himself or for another person.

R.R.Q., 1981, c. I-9, r. 3, s. 3.06.03.

3.06.04. An engineer shall not accept a mandate which entails or may entail the disclosure or use of confidential information or documents obtained from another client without the latter's consent.

R.R.Q., 1981, c. I-9, r. 3, s. 3.06.04.

§7. Access to and correction of records and release of documents

R.R.Q., 1981, c. I-9, r. 3, sd. 7; O.C. 920-2002, s. 1.

3.07.01. Beyond the specific rules prescribed by law, an engineer must act, with diligence and not later than 30 days following receipt thereof, on any request made by his client for the purposes of:

(1) examining documents concerning him in any record established in his respect;

(2) obtaining copies of documents concerning him in any record established in his respect.

R.R.Q., 1981, c. I-9, r. 3, s. 3.07.01; O.C. 920-2002, s. 1.

3.07.02. An engineer who agrees to a request contemplated in section 3.07.01 shall give the client access to the documents in his presence or in the presence of a person authorized by him.

An engineer may, with respect to a request contemplated in subparagraph 2 of section 3.07.01, charge his client a reasonable fee not exceeding the cost of transmission, transcription or reproduction of a copy.

An engineer charging such fees shall, before they are incurred, inform his client of the approximate amount he will be asked to pay. An engineer has the right of retention concerning payment of such fees.

O.C. 920-2002, s. 1.

3.07.03. An engineer who, in applying the second paragraph of section 60.5 of the Professional Code (chapter C-26), refuses to allow his client access to information contained in any record established in his respect, must furnish his client with the reasons for such refusal in writing.

O.C. 920-2002, s. 1.

3.07.04. Beyond the specific rules prescribed by law, an engineer must act, with diligence and not later than 30 days following receipt thereof, on any request made by his client for the purposes of:

(1) correcting information that is inaccurate, incomplete or ambiguous with regard to the purposes for which it was collected, in any document concerning him that is contained in any record established in his respect;

(2) deleting any information that is outdated or not justified by the object of the record established in his respect;

(3) placing his written comments in the record established in his respect.

O.C. 920-2002, s. 1.

3.07.05. An engineer who agrees to a request contemplated in section 3.07.04 shall give his client without charge a copy of the document or portion thereof showing the client that the information has been corrected, or, as the case may be, a certificate indicating that the written comments from the client have been placed in the record.

Upon receipt of a request in writing from the client, an engineer shall send, without charge to the client, a copy of such information or certificate to any person from whom an engineer received such information and to whom such information was given.

O.C. 920-2002, s. 1.

3.07.06. An engineer agrees to act with diligence on any request in writing made by his client for the purpose of taking back a document or item which the client had left with him.

The engineer indicates in the record established in respect of his client, as the case may be, the reasons for the client's request.

O.C. 920-2002, s. 1.

3.07.07. An engineer may require that a request contemplated in section 3.07.01, 3.07.04 or 3.07.06 be submitted to his professional domicile during the usual hours of work.

O.C. 920-2002, s. 1.

§8. Determination and payment of fees

3.08.01. An engineer must charge and accept fair and reasonable fees.

R.R.Q., 1981, c. I-9, r. 3, s. 3.08.01.

3.08.02. Fees are considered fair and reasonable when they are justified by the circumstances and correspond to the services rendered. In determining his fees, the engineer must, in particular, take the following factors into account:

(a) the time devoted to the carrying out of the mandate;

(b) the difficulty and magnitude of the mandate;

(c) the performance of unusual services or services requiring exceptional competence or speed;

(d) the responsibility assumed.

R.R.Q., 1981, c. I-9, r. 3, s. 3.08.02.

3.08.03. An engineer must inform his client of the approximate cost of his services and of the terms and conditions of payment. He must refrain from demanding advance payment of his fees; he may, however, request a deposit.

R.R.Q., 1981, c. I-9, r. 3, s. 3.08.03; O.C. 2566-84, s. 5.

3.08.04. An engineer must give his client all the necessary explanations for the understanding of his statement of fees and the terms and conditions of its payment.

R.R.Q., 1981, c. I-9, r. 3, s. 3.08.04.

DIVISION IV
DUTIES AND OBLIGATIONS TOWARDS THE PROFESSION

§1. Derogatory acts

4.01.01. In addition to those referred to in sections 57 and 58 of the Professional Code (chapter C-26), the following acts are derogatory to the dignity of the profession:

(a) participating or contributing to the illegal practice of the profession;

(b) pressing or repeated inducement to make use of his professional services;

(c) communicating with the person who lodged a complaint, without the prior written permission of the syndic or his assistant, whenever he is informed of an inquiry into his professional conduct or competence or whenever a complaint has been laid against him;

(d) refusing to comply with the procedures for the conciliation and arbitration of accounts and with the arbitrators' award;

(e) taking legal action against a colleague on a matter relative to the practice of the profession before applying for conciliation to the president of the Order;

(f) refusing or failing to present himself at the office of the syndic, of one of his assistants or of a corresponding syndic, upon request to that effect by one of those persons;

(g) not notifying the syndic without delay if he believes that an engineer infringes this Regulation.

R.R.Q., 1981, c. I-9, r. 3, s. 4.01.01.

§2. *Relations with the Order and colleagues*

4.02.01. An engineer whose participation in a council for the arbitration of accounts, a disciplinary council or a professional inspection committee is requested by the Order, must accept this duty unless he has exceptional grounds for refusing.

R.R.Q., 1981, c. I-9, r. 3, s. 4.02.01.

4.02.02. An engineer must, as promptly as possible, answer all correspondence addressed to him by the syndic of the Order, the assistant syndic or a corresponding syndic, investigators or members of the professional inspection committee or the secretary of the latter committee.

R.R.Q., 1981, c. I-9, r. 3, s. 4.02.02.

4.02.03. An engineer shall not abuse a colleague's good faith, be guilty of breach of trust or be disloyal towards him or willfully damage his reputation. Without restricting the generality of the foregoing, the engineer shall not, in particular:

(a) take upon himself the credit for engineering work which belongs to a colleague;

(b) take advantage of his capacity of employer or executive to limit in any way the professional independence of an engineer employed by him or under his responsibility, in particular with respect to the use of the title of engineer or the obligation of every engineer to commit his professional liability;

(c) induce a colleague to commit an offence against the laws and regulations governing the practice of the profession.

R.R.Q., 1981, c. I-9, r. 3, s. 4.02.03; O.C. 2566-84, s. 6.

4.02.04. Where a client requests an engineer to examine or review engineering work that he has not performed himself, the latter must notify the engineer concerned thereof and, where applicable, ensure that the mandate of his colleague has terminated.

R.R.Q., 1981, c. I-9, r. 3, s. 4.02.04.

4.02.05. Where an engineer replaces a colleague in engineering work, he must notify that colleague thereof and make sure that the latter's mandate has terminated.

R.R.Q., 1981, c. I-9, r. 3, s. 4.02.05.

4.02.06. An engineer who is called upon to collaborate with a colleague must retain his professional independence. If a task is entrusted to him and such task goes against his conscience or his principles, he may ask to be excused from doing it.

R.R.Q., 1981, c. I-9, r. 3, s. 4.02.06.

4.02.07. An engineer may not refuse to collaborate with a member of the Order, in professional dealings, on the basis of race, colour, sex, religion, national, ethnic or social origin and for any ground mentioned in section 10 of the Charter of human rights and freedoms (chapter C-12).

O.C. 1182-83, s. 1.

§3. Contribution to the advancement of the profession

4.03.01. An engineer must, as far as he is able, contribute to the development of his profession by sharing his knowledge and experience with his colleagues and students, and by his participation as professor or tutor in continuing training periods and refresher training courses.

R.R.Q., 1981, c. I-9, r. 3, s. 4.03.01.

DIVISION V
OBLIGATIONS RELATIVE TO PROFESSIONAL ADVERTISING AND PROMOTION AND OBLIGATIONS RELATIVE TO THE NAMES OF PARTNERSHIPS OF ENGINEERS

O.C.920-2002,s.2.

§1. Advertising and promotion

O.C.920-2002,s.2.

5.01.01. An engineer may not in any way and under any circumstances make false, misleading or incomplete advertising with respect to his professional activities and services.

O.C. 920-2002, s. 2.

5.01.02. The information that an engineer provides in his advertising or promotion must be of a nature to help the public make an informed choice. Such advertising or promotion must be done with integrity and favour professionalism.

O.C. 920-2002, s. 2.

5.01.03. In all advertising or representation he may make, an engineer must give his name and professional title.

O.C. 920-2002, s. 2.

5.01.04. An engineer shall not in his representation or advertising:

(1) invade a person's privacy;

(2) undermine a person's reputation;

(3) compare the quality of his services with that of the services offered or rendered by other engineers;

(4) discredit, denigrate or disparage the services offered or rendered by other engineers.

O.C. 920-2002, s. 2.

5.01.05. In addition to the obligations mentioned in section 5.01.04, an engineer shall not attribute to himself experience, professional or academic qualifications or particular qualities unless he is able to justify them.

O.C. 920-2002, s. 2.

5.01.06. An engineer shall ensure that the persons working with him in any capacity in the practice of his profession comply with the rules concerning advertising.

O.C. 920-2002, s. 2.

5.01.07. An engineer who, in his advertising, mentions fees or prices shall do so in a manner that can be understood by the public, which has no particular knowledge of the practice of engineering or the professional services covered by the advertising, and shall:

(1) keep them in effect for the period mentioned in the advertising or, if no period is specified, for a period of 90 days following the last publication or broadcast;

(2) specify the nature and extent of the services included in such fees or prices;

(3) indicate whether or not certain fees are included in such fees or prices;

(4) indicate what additional services may be required which are not included in such fees or prices.

O.C. 920-2002, s. 2.

5.01.08. In the case of advertising offering a special price or a discount, an engineer shall specify how long such special price or discount is valid, as the case may be. This period may be less than 90 days.

O.C. 920-2002, s. 2.

5.01.09. An engineer shall keep a copy of all advertising for a period of 3 years following the date of its last broadcast or publication. On request, this copy shall be given to the syndic.

O.C. 920-2002, s. 2.

§2. Names of partnerships of engineers

O.C.920-2002,s.2.

5.02.01. The name of a partnership of engineers includes only the names of the engineers who are practising their profession together. It may not include the name of a deceased or retired associate engineer for more than one year, unless he or his successors had made an agreement in writing to the contrary.

O.C. 920-2002, s. 2.

5.02.02. When an associate engineer withdraws from a partnership to practise alone, to join another partnership or another business or to hold a position that is incompatible with the practice of the profession, his name must be eliminated from the name within 30 days of his withdrawal, unless there is a written agreement to the contrary.

In all cases, the agreement may not stipulate a period of more than one year.

O.C. 920-2002, s. 2.

5.02.03. The name of a partnership of engineers may end with the words "and associates" when the names of at least 2 associates are not included in the name.

O.C. 920-2002, s. 2.

5.02.04. An engineer practising in a partnership is jointly responsible with the other professionals for following the rules concerning advertising, unless he can establish that the advertising was done without his knowledge or consent and in spite of the provisions made to ensure compliance with such rules.

O.C. 920-2002, s. 2.

1-3 ACCIDENT AND LOSS STATISTICS

Accident and loss statistics are important measures of the effectiveness of safety programs. These statistics are valuable for determining whether a process is safe or whether a safety procedure is working effectively.

Many statistical methods are available to characterize accident and loss performance. These statistics must be used carefully. Like most statistics they are only averages and do not reflect the potential for single episodes involving substantial losses. Unfortunately, no single method is capable of measuring all required aspects. The three systems considered here are

- OSHA incidence rate,
- Fatal accident rate (FAR), and

• Fatality rate, or deaths per person per year

All three methods report the number of accidents and/or fatalities for a fixed number of workers during a specified period.

OSHA stands for the Occupational Safety and Health Administration of the United States government. OSHA is responsible for ensuring that workers are provided with a safe working environment. Table 1-2 contains several OSHA definitions applicable to accident statistics.

Table 1-2. Glossary of Terms Used by OSHA and Industry to Represent Work-Related Losses

Term	Definition
First aid	Any one-time treatment and any follow-up visits for the purpose of observation of minor scratches, cuts, burns, splinters, and so forth that do not ordinarily require medical care. Such one-time treatment and follow-up visits for the purpose of observation are considered first aid even though provided by a physician or registered professional personnel.
Incident rate	Number of occupational injuries and/or illnesses or lost workdays per 100 full-time employees.
Lost workdays	Number of days (consecutive or not) after but not including the day of injury or illness during which the employee would have worked but could not do so, that is, during which the employee could not perform all or any part of his or her normal assignment during all or any part of the workday or shift because of the occupational injury or illness.
Medical treatment	Treatment administered by a physician or by registered professional personnel under the standing orders of a physician. Medical treatment does not include first aid treatment even though provided by a physician or registered professional personnel.
Occupational injury	Any injury such as a cut, sprain, or burn that results from a work accident or from a single instantaneous exposure in the work environment.
Occupational illness	Any abnormal condition or disorder, other than one resulting from an occupational injury, caused by exposure to environmental factors associated with employment. It includes acute and chronic illnesses or diseases that may be caused by inhalation, absorption, ingestion, or direct contact.
Recordable cases	Cases involving an occupational injury or occupational illness, including deaths.
Recordable fatality cases	Injuries that result in death, regardless of the time between the injury and death or the length of the illness.

Term	Definition
Recordable nonfatal cases without lost workdays	Cases of occupational injury or illness that do not involve fatalities or lost workdays but do result in (1) transfer to another job or termination of employment or (2) medical treatment other than first aid or (3) diagnosis of occupational illness or (4) loss of consciousness or (5) restriction of work or motion.
Recordable lost workday cases due to restricted duty	Injuries that result in the injured person not being able to perform their regular duties but being able to perform duties consistent with their normal work.
Recordable cases with days away from work	Injuries that result in the injured person not being able to return to work on their next regular workday.
Recordable medical cases	Injuries that require treatment that must be administered by a physician or under the standing orders of a physician. The injured person is able to return to work and perform his or her regular duties. Medical injuries include cuts requiring stitches, second-degree burns (burns with blisters), broken bones, injury requiring prescription medication, and injury with loss of consciousness.

The OSHA incidence rate is based on cases per 100 worker years. A worker year is assumed to contain 2000 hours (50 work weeks/year x 40 hours/week). The OSHA incidence rate is therefore based on 200,000 hours of worker exposure to a hazard. The OSHA incidence rate is calculated from the number of occupational injuries and illnesses and the total number of employee hours worked during the applicable period. The following equation is used:

Equation 1-1

$$\text{OSHA incidence rate (based on injuries and illness)} = \frac{\text{Number of injuries and illnesses} \times 200{,}000}{\text{Total hours worked by all employees during period covered.}}$$

An incidence rate can also be based on lost workdays instead of injuries and illnesses. For this case

Equation 1-2

$$\text{OSHA incidence rate (based on lost workdays)} = \frac{\text{Number of lost workdays} \times 200{,}000}{\text{Total hours worked by all employees during period covered.}}$$

The definition of a lost workday is given in Table 1-2.

The OSHA incidence rate provides information on all types of work-related injuries and illnesses, including fatalities. This provides a better representation of worker accidents than systems based on fatalities alone. For instance, a plant might experience many small accidents with resulting injuries but no fatalities. On the other hand, fatality data cannot be extracted from the OSHA incidence rate without additional information.

The FAR is used mostly by the British chemical industry. This statistic is used here because there are some useful and interesting FAR data available in the open literature. The FAR reports the number of fatalities based on 1000 employees working their entire lifetime. The employees are assumed to work a total of 50 years. Thus the FAR is based on 10^8 working hours. The resulting equation is

Equation 1-3

$$FAR = \frac{\text{Number of fatalities} \times 10^8}{\text{Total hours worked by all employees during period covered.}}$$

The last method considered is the fatality rate or deaths per person per year. This system is independent of the number of hours actually worked and reports only the number of fatalities expected per person per year. This approach is useful for performing calculations on the general population, where the number of exposed hours is poorly defined. The applicable equation is

Equation 1-4

$$\text{Fatality rate} = \frac{\text{Number of fatalities per year}}{\text{Total number of people in applicable population.}}$$

Both the OSHA incidence rate and the FAR depend on the number of exposed hours. An employee working a ten-hour shift is at greater total risk than one working an eight-hour shift. A FAR can be converted to a fatality rate (or vice versa) if the number of exposed hours is known. The OSHA incidence rate cannot be readily converted to a FAR or fatality rate because it contains both injury and fatality information.

Example 1-1.

A process has a reported FAR of 2. If an employee works a standard 8-hr shift 300 days per year, compute the deaths per person per year.

Solution

Deaths per person per year	=	(8 hr/day) x (300 days/yr) x (2 deaths/10^8 hr)
	=	4.8×10^{-5}.

Typical accident statistics for various industries are shown in Table 1-3. A FAR of 1.2 is reported in Table 1-3 for the chemical industry. Approximately half these deaths are due to ordinary industrial accidents (falling down stairs, being run over), the other half to chemical exposures.[2]

Table 1-3. Accident Statistics for Selected Industries

	OSHA incident rates (U.S.)					
	Recordable	Days Away from Work	Fatality		FAR (UK)[c]	
Industrial activity	2007	2007	2000	2005	1974–78	1987–90
Agriculture[1]	6.1	3.2	24.1	27	7.4	3.7
Chemical and allied products	3.3	1.9	2.5	2.8	2.4	1.2
Coal mining	4.7	3.2	50	26.8	14.5	7.3
Construction	5.4	2.8	10	11.1	10	5.0
Vehicle manufacturing	9.3	5.0	1.3	1.7	1.2	0.6
All manufacturing	5.6	3.0	3.3	2.4	2.3	1.2

The FAR figures show that if 1000 workers begin employment in the chemical industry, 2 of the workers will die as a result of their employment throughout all of their working lifetimes. One of these deaths will be due to direct chemical exposure. However, 20 of these same 1000 people will die as a result of nonindustrial accidents (mostly at home or on the road) and 370 will die from disease. Of those that perish from disease, 40 will die as a direct result of smoking.[3]

Table 1-4 lists the FARs for various common activities. The table is divided into voluntary and involuntary risks. Based on these data, it appears that individuals are willing to take a substantially greater risk if it is voluntary. It is also evident that many common everyday activities are substantially more dangerous than working in a chemical plant.

Table 1-4. Fatality Statistics for Common Nonindustrial Activities

Activity	FAR (deaths/10^8 hours)	Fatality rate (deaths per person per year)
Voluntary activity		
Staying at home	3	
Traveling by		

Activity	FAR (deaths/10^8hours)	Fatality rate (deaths per person per year)
Car	57	17×10^{-5}
Bicycle	96	
Air	240	
Motorcycle	660	
Canoeing	1000	
Rock climbing	4000	4×10^{-5}
Smoking (20 ciga-rettes/day)		500×10^{-5}
Involuntary activity		
Struck by meteorite		6×10^{-11}
Struck by lightning (U.K.)		1×10^{-7}
Fire (U.K.)		150×10^{-7}
Run over by vehicle		600×10^{-7}

For example, Table 1-4 indicates that canoeing is much more dangerous than traveling by motorcycle, despite general perceptions otherwise. This phenomenon is due to the number of exposed hours. Canoeing produces more fatalities per hour of activity than traveling by motorcycle. The total number of motorcycle fatalities is larger because more people travel by motorcycle than canoe.

Example 1-2.

If twice as many people used motorcycles for the same average amount of time each, what will happen to (a) the OSHA incidence rate, (b) the FAR, (c) the fatality rate, and (d) the total number of fatalities?

Solution

a. The OSHA incidence rate will remain the same. The number of injuries and deaths will double, but the total number of hours exposed will double as well.
b. The FAR will remain unchanged for the same reason as in part a.
c. The fatality rate, or deaths per person per year, will double. The fatality rate does not depend on exposed hours.
d. The total number of fatalities will double.

Example 1-3.

If all riders used their motorcycles twice as much, what will happen to (a) the OSHA incidence rate, (b) the FAR, (c) the fatality rate, and (d) the total number of fatalities?

Solution

   a. The OSHA incidence rate will remain the same. The same reasoning applies as for Example 1-2, part a.

   b. The FAR will remain unchanged for the same reason as in part a.

   c. The fatality rate will double. Twice as many fatalities will occur within this group.

   d. The number of fatalities will double.

Example 1-4.

A friend states that more rock climbers are killed traveling by automobile than are killed rock climbing. Is this statement supported by the accident statistics?

Solution

The data from Table 1-4 show that traveling by car (FAR = 57) is safer than rock climbing (FAR = 4000). Rock climbing produces many more fatalities per exposed hour than traveling by car. However, the rock climbers probably spend more time traveling by car than rock climbing. As a result, the statement might be correct but more data are required.

Recognizing that the chemical industry is safe, why is there so much concern about chemical plant safety? The concern has to do with the industry's potential for many deaths, as, for example, in the Bhopal, India, tragedy. Accident statistics do not include information on the total number of deaths from a single incident. Accident statistics can be somewhat misleading in this respect. For example, consider two separate chemical plants. Both plants have a probability of explosion and complete devastation once every 1000 years. The first plant employs a single operator. When the plant explodes, the operator is the sole fatality. The second plant employs 10 operators. When this plant explodes all 10 operators succumb. In both cases the FAR and OSHA incidence rate are the same; the second accident kills more people, but there are a correspondingly larger number of exposed hours. In both cases the risk taken by an individual operator is the same.

It is human nature to perceive the accident with the greater loss of life as the greater tragedy. The potential for large loss of life gives the perception that the chemical industry is unsafe.

Loss data published for losses after 1966 and in 10-year increments indicate that the total number of losses, the total dollar amount lost, and the average amount lost per incident have steadily increased. The total loss figure has doubled every 10 years despite increased efforts by the chemical process industry to improve safety. The increases are mostly due to an expansion in the number of chemical plants, an increase in chemical plant size, and an increase in the use of more complicated and dangerous chemicals.

Property damage and loss of production must also be considered in loss prevention. These losses can be substantial. Accidents of this type are much more common than fatalities. This is demonstrated in the accident pyramid. The numbers provided are only approximate. The exact numbers vary by industry, location, and time. "No Damage" accidents are frequently called "near misses" and provide a good opportunity for companies to determine that a problem exists and to correct it before a more serious accident occurs. It is frequently said that "the cause of an accident is visible the day before it occurs." Inspections, safety reviews, and careful evaluation of near misses will identify hazardous conditions that can be corrected before real accidents occur.

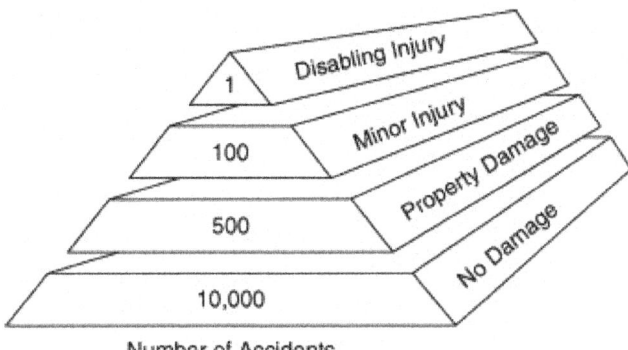

Number of Accidents

Safety is good business and, like most business situations, has an optimal level of activity beyond which there are diminishing returns. As shown by Kletz, if initial expenditures are made on safety, plants are prevented from blowing up and experienced workers are spared. This results in increased return because of reduced loss expenditures. If safety expenditures increase, then the return increases more, but it may not be as much as before and not as much as achieved by spending money elsewhere. If safety expenditures increase further, the price of the product increases and sales diminish. Indeed, people are spared from injury (good humanity), but the cost is decreased sales. Finally, even higher safety expenditures result in uncompetitive product pricing: The company will go out of business. Each company needs to determine an appropriate level for safety expenditures. This is part of risk management.

From a technical viewpoint, excessive expenditures for safety equipment to solve single safety problems may make the system unduly complex and consequently may cause new safety problems because of this complexity. This excessive expense could have a higher safety return if assigned to a different safety problem. Engineers need to also consider other alternatives when designing safety improvements.

It is also important to recognize the causes of accidental deaths, as shown in Table 1-5. Because most, if not all, company safety programs are directed toward preventing injuries to employees, the programs should include off-the-job safety, especially training to prevent accidents with motor vehicles.

Table 1-5. All Accidental Deaths

Type of death	1998 deaths	2007 deaths
Motor-vehicle		
Public nonwork	38,900	40,955
Work	2,100	1,945
Home	200	200
Subtotal	41,200 (43.5%)	43,100 (35.4%)
Work		
Non-motor-vehicle	3,000	2,744
Motor-vehicle	2,100	1,945
Subtotal	5,100 (5.4%)	4,689 (3.9%)
Home		
Non-motor-vehicle	28,200	43,300
Motor-vehicle	200	200
Subtotal	28,400 (30.0%)	43,500 (35.7%)
Public	20,000 (21.1%)	30,500 (25%)
All classes	94,700	121,789

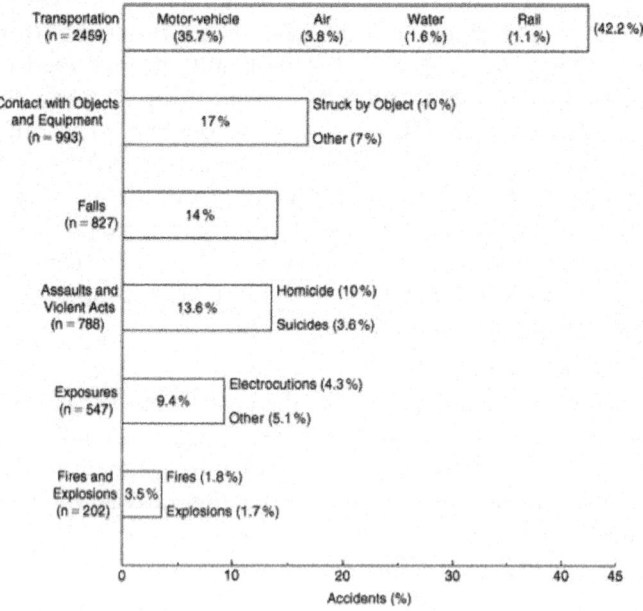

Figure. The manner in which workplace fatalities occurred in 2006. The total number of workplace fatalities was 5840; this includes the above plus 14 for bodily reaction and exertion, and 10 nonclassified.

When organizations focus on the root causes of worker injuries, it is helpful to analyze the manner in which workplace fatalities occur (see Figure 1-4). Although the emphasis of this book is the prevention of chemical-related accidents, the data in Figure 1-4 show that safety programs need to include training to prevent injuries resulting from transportation, assaults, mechanical and chemical exposures, and fires and explosions.

1-4 Acceptable Risk

We cannot eliminate risk entirely. Every chemical process has a certain amount of risk associated with it. At some point in the design stage someone needs to decide if the risks are "acceptable." That is, are the risks greater than the normal day-to-day risks taken by individuals in their nonindustrial environment? Certainly it would require a substantial effort and considerable expense to design a process with a risk comparable to being struck by lightning. Is it satisfactory to design a process with a risk comparable to the risk of sitting at home? For a single chemical process in a plant composed of several processes, this risk may be too high because the risks resulting from multiple exposures are additive.[7]

Engineers must make every effort to minimize risks within the economic constraints of the process. No engineer should ever design a process that he or she knows will result in certain human loss or injury, despite any statistics.

In the field of engineering, a **chemical engineer** is a professional who works principally in the chemical industry to convert basic raw materials into a variety of products, and deals with the design and operation of plants and equipment to perform such work.[1] In general, a chemical engineer is one who applies and uses principles of chemical engineering in any of its various practical applications; these often include 1) design, manufacture, and operation of plants and machinery in industrial chemical and related processes ("chemical process engineers"); 2) development of new or adapted substances for products ranging from foods and beverages to cosmetics to cleaners to pharmaceutical ingredients, among many other products ("chemical product engineers"); and 3) development of new technologies such as fuel cells, hydrogen power and nanotechnology, as well as working in fields wholly or partially derived from Chemical Engineering such as materials science, polymer engineering, and biomedical engineering.

1-5 Public Perceptions

The general public has great difficulty with the concept of acceptable risk. The major objection is due to the involuntary nature of acceptable risk. Chemical plant designers who specify the acceptable risk are assuming that these risks are satisfactory to the civilians living near the plant. Frequently these civilians are not aware that there is any risk at all.

The results of a public opinion survey on the hazards of chemicals are shown in Figure 1-5. This survey asked the participants if they would say chemicals do more good than harm, more harm than good, or about the same amount of each.

The results show an almost even three-way split, with a small margin to those who considered the good and harm to be equal.

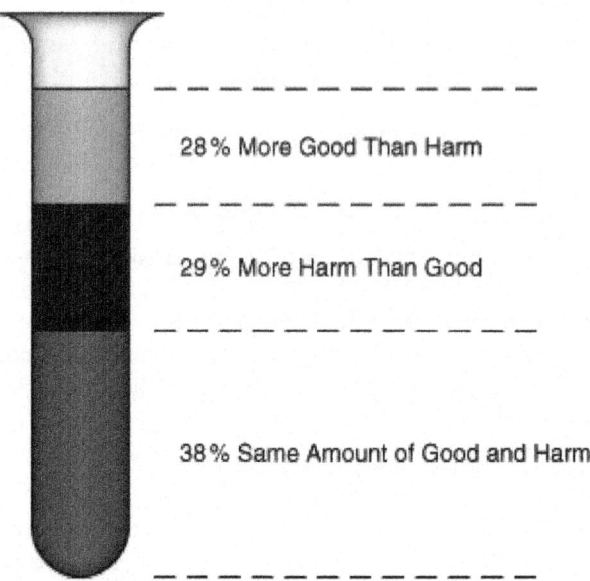

Figure 1-5 Results from a public opinion survey asking the question, "Would you say chemicals do more good than harm, more harm than good, or about the same amount of each?" Source: The Detroit News.

Some naturalists suggest eliminating chemical plant hazards by "returning to nature." One alternative, for example, is to eliminate synthetic fibers produced by chemicals and use natural fibers such as cotton. As suggested by Kletz, accident statistics demonstrate that this will result in a greater number of fatalities because the FAR for agriculture is higher.

Example 1-5.

List six different products produced by chemical engineers that are of significant benefit to mankind.

Solution

Penicillin, gasoline, synthetic rubber, paper, plastic, concrete.

1-6 The Nature of the Accident Process

Chemical plant accidents follow typical patterns. It is important to study these patterns in order to anticipate the types of accidents that will occur. As shown in Table 1-6, fires are the most common, followed by explosion and toxic release. With respect to fatalities, the order reverses, with toxic release having the greatest potential for fatalities.

Table 1-6. Three Types of Chemical Plant Accidents

Type of accident	Probability of occurrence	Potential for fatalities	Potential for economic loss
Fire	High	Low	Intermediate
Explosion	Intermediate	Intermediate	High
Toxic release	Low	High	Low

Economic loss is consistently high for accidents involving explosions. The most damaging type of explosion is an unconfined vapor cloud explosion, where a large cloud of volatile and flammable vapor is released and dispersed throughout the plant site followed by ignition and explosion of the cloud. An analysis of the largest chemical plant accidents (based on worldwide accidents and 1998 dollars) is provided in Figure. As illustrated, vapor cloud explosions account for the largest percentage of these large losses. The "other" category of Figure includes losses resulting from floods and windstorms.

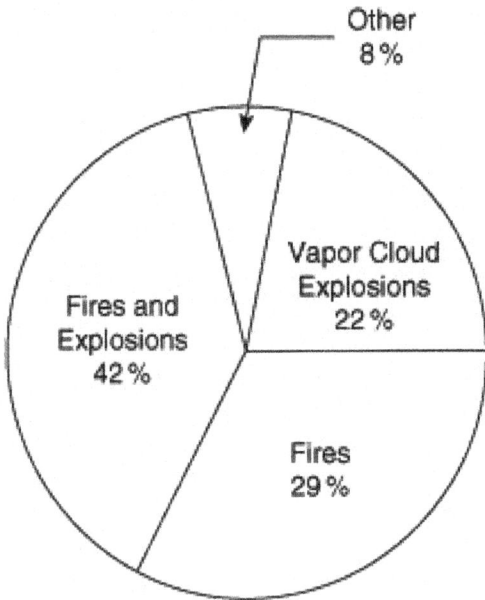

Figure Types of loss for large hydrocarbon-chemical plant accidents. Data from the 100 Largest Losses, 1972-2001.

Toxic release typically results in little damage to capital equipment. Personnel injuries, employee losses, legal compensation, and cleanup liabilities can be significant.

Presents the causes of losses for these largest accidents. By far the most frequent cause is mechanical failures, such as pipe failures due to corrosion, erosion, and high pressures, and seal/gasket failures. Failures of this type are usually due to poor maintenance or the poor utilization of the principles of inherent safety (Section 1-7) and process safety management (Section 3-1). Pumps, valves, and control equipment will fail if not properly maintained. The second largest cause is operator error. For example, valves are not opened or closed in the proper sequence or reactants are not charged to a reactor in the correct order. Process upsets caused by, for example, power or cooling water failures account for 3% of the losses.

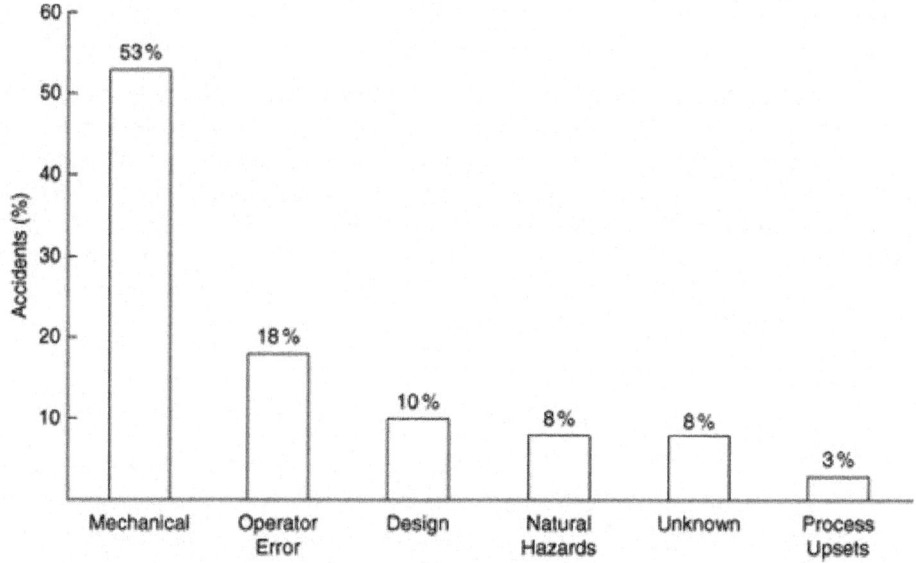

Causes of losses for largest hydrocarbon-chemical plant accidents. Data from the 100 Largest Losses, 1972-2001.

Human error is frequently used to describe a cause of losses. Almost all accidents, except those caused by natural hazards, can be attributed to human error. For instance, mechanical failures could all be due to human error as a result of improper maintenance or inspection. The term "operator error," used in, includes human errors made on-site that led directly to the loss.

Presents a survey of the type of hardware associated with large accidents. Piping system failure represents the bulk of the accidents, followed by storage tanks and reactors. An interesting result of this study is that the most complicated mechanical components (pumps and compressors) are minimally responsible for large losses.

The loss distribution for the hydrocarbon and chemical industry over 5-year intervals. The number and magnitude of the losses increase over each consecutive 10-year period for the past 30 years. This increase corresponds to the trend of building larger and more complex plants.

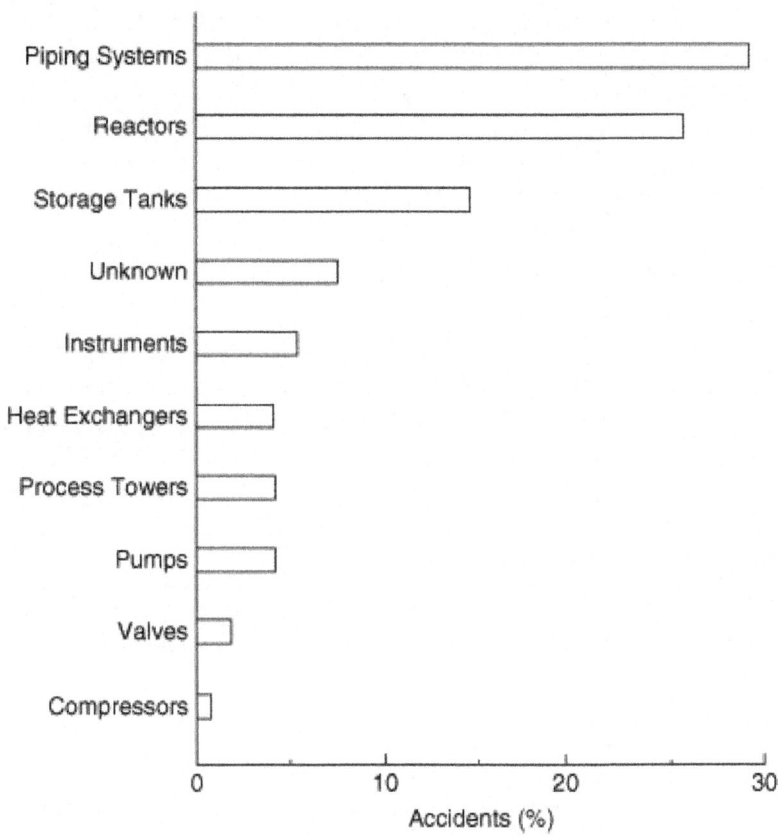

Hardware associated with the largest hydrocarbon-chemical plant accidents.
Data from The 100 Largest Losses, 1972-2001.

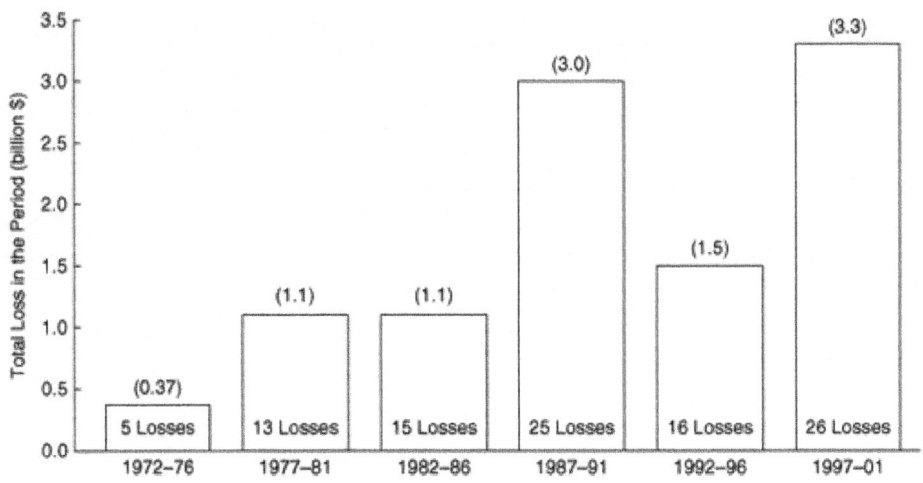

Loss distribution for the largest hydrocarbon-chemical plant accidents over a
30-year period. Data from The 100 Largest Losses, 1972-2001.

The lower losses between 1992 and 1996 are likely the temporary result of governmental regulations that were implemented in the United States during this time; that is, on February 24, 1992, OSHA published its final rule "Process Safety Management of Highly Hazardous Chemicals (PSM)." This rule became effective on May 26, 1992. As shown, however, the lower losses between 1992 and 1996 were probably a start-up benefit of PSM because in the last 5-year period (1997-01) the losses went up again.

Accidents follow a three-step process. The following chemical plant accident illustrates these steps.

A worker walking across a high walkway in a process plant stumbles and falls toward the edge. To prevent the fall, he grabs a nearby valve stem. Unfortunately, the valve stem shears off and flammable liquid begins to spew out. A cloud of flammable vapor rapidly forms and is ignited by a nearby truck. The explosion and fire quickly spread to nearby equipment. The resulting fire lasts for six days until all flammable materials in the plant are consumed, and the plant is completely destroyed.

This disaster occurred in 1969 and led to an economic loss of $4,161,000. It demonstrates an important point: Even the simplest accident can result in a major catastrophe.

Most accidents follow a three-step sequence:

- Initiation (the event that starts the accident),
- Propagation (the event or events that maintain or expand the accident), and
- Termination (the event or events that stop the accident or diminish it in size).

In the example the worker tripped to initiate the accident. The accident was propagated by the shearing of the valve and the resulting explosion and growing fire. The event was terminated by consumption of all flammable materials.

Safety engineering involves eliminating the initiating step and replacing the propagation steps with termination events. Table 1-7 presents a few ways to accomplish this. In theory, accidents can be stopped by eliminating the initiating step. In practice this is not effective: It is unrealistic to expect elimination of all initiations. A much more effective approach is to work on all three areas to ensure that accidents, once initiated, do not propagate and will terminate as quickly as possible.

Table 1-7. Defeating the Accident Process

Step	Desired effect	Procedure
Initiation	Diminish	Grounding and bonding Inerting Explosion proof electrical Guardrails and guards Maintenance procedures Hot work permits Human factors design Process design Awareness of dangerous properties of chemicals
Propagation	Diminish	Emergency material transfer Reduce inventories of flammable materials Equipment spacing and layout Nonflammable construction materials Installation of check and emergency shutoff valves
Termination	Increase	Fire-fighting equipment and procedures Relief systems Sprinkler systems Installation of check and emergency shutoff valves

Example 1-6.

The following accident report has been filed

- Failure of a threaded 1 1/2" drain connection on a rich oil line at the base of an absorber tower in a large (1.35 MCF/D) gas producing plant allowed the release of rich oil and gas at 850 psi and –40°F. The resulting vapor cloud probably ignited from the ignition system of engine-driven recompressors. The 75′ high x 10′ diameter absorber tower eventually collapsed across the pipe rack and on two exchanger trains. Breaking pipelines added more fuel to the fire. Severe flame impingement on an 11,000-horsepower gas turbine–driven compressor, waste heat recovery, and super-heater train resulted in its near total destruction.

Identify the initiation, propagation, and termination steps for this accident.

Solution

Initiation:	Failure of threaded 1 1/2" drain connection
Propagation:	Release of rich oil and gas, formation of vapor cloud, ignition of vapor cloud by recompressors, collapse of absorber tower across pipe rack
Termination:	Consumption of combustible materials in process

As mentioned previously, the study of case histories is an especially important step in the process of accident prevention. To understand these histories, it is helpful to know the definitions of terms that are commonly used in the descriptions (see Table 1-8).

Table 1-8. Definitions for Case Histories

Term	Definition
Accident	The occurrence of a sequence of events that produce unintended injury, death, or property damage. "Accident" refers to the event, not the result of the event.
Hazard	A chemical or physical condition that has the potential for causing damage to people, property, or the environment.
Incident	The loss of containment of material or energy; not all events propagate into incidents; not all incidents propagate into accidents.
Consequence	A measure of the expected effects of the results of an incident.
Likelihood	A measure of the expected probability or frequency of occurrence of an event. This may be expressed as a frequency, a probability of occurrence during some time interval, or a conditional probability.
Risk	A measure of human injury, environmental damage, or economic loss in terms of both the incident likelihood and the magnitude of the loss or injury.
Risk analysis	The development of a quantitative estimate of risk based on an engineering evaluation and mathematical techniques for combining estimates of incident consequences and frequencies.
Risk assessment	The process by which the results of a risk analysis are used to make decisions, either through a relative ranking of risk reduction strategies or through comparison with risk targets.
Scenario	A description of the events that result in an accident or incident. The description should contain information relevant to defining the root causes.

1-7 Inherent Safety

An inherently safe plant relies on chemistry and physics to prevent accidents rather than on control systems, interlocks, redundancy, and special operating procedures to prevent accidents. Inherently safer plants are tolerant of errors and are often the most cost effective. A process that does not require complex safety interlocks and elaborate procedures is simpler, easier to operate, and more reliable. Smaller equipment, operated at less severe temperatures and pressures, has lower capital and operating costs.

In general, the safety of a process relies on multiple layers of protection. The first layer of protection is the process design features. Subsequent layers include control systems, interlocks, safety shutdown systems, protective systems, alarms, and emergency response plans. Inherent safety is a part of all layers of protection; however, it is especially directed toward process design features. The best approach to prevent accidents is to add process design features to prevent hazardous situations. An inherently safer plant is more tolerant of operator errors and abnormal conditions.

Although a process or plant can be modified to increase inherent safety at any time in its life cycle, the potential for major improvements is the greatest at the earliest stages of process development. At these early stages process engineers and chemists have the maximum degree of freedom in the plant and process specifications, and they are free to consider basic process alternatives, such as changes to the fundamental chemistry and technology.

The following four words are recommended to describe inherent safety:

- Minimize (intensification)
- Substitute (substitution)
- Moderate (attenuation and limitation of effects)
- Simplify (simplification and error tolerance).

The types of inherent safety techniques that are used in the chemical industry are illustrated in Table 1-9 and are described more fully in what follows.

Table 1-9. Inherent Safety Techniques

Type	Typical techniques
Minimize (intensification)	Change from large batch reactor to a smaller continuous reactor Reduce storage inventory of raw materials Improve control to reduce inventory of hazardous intermediate chemicals Reduce process hold-up
Substitute (substitution)	Use mechanical pump seals *vs.* packing Use welded pipe *vs.* flanged Use solvents that are less toxic Use mechanical gauges *vs.* mercury Use chemicals with higher flash points, boiling points, and other less hazardous properties Use water as a heat transfer fluid instead of hot oil
Moderate (attenuation and limitation of effects)	Use vacuum to reduce boiling point Reduce process temperatures and pressures Refrigerate storage vessels Dissolve hazardous material in safe solvent Operate at conditions where reactor runaway is not possible Place control rooms away from operations Separate pump rooms from other rooms Acoustically insulate noisy lines and equipment Barricade control rooms and tanks

Type	Typical techniques
Simplify (simplification and error tolerance)	Keep piping systems neat and visually easy to follow
	Design control panels that are easy to comprehend
	Design plants for easy and safe maintenance
	Pick equipment that requires less maintenance
	Pick equipment with low failure rates
	Add fire- and explosion-resistant barricades
	Separate systems and controls into blocks that are easy to comprehend and understand
	Label pipes for easy "walking the line"
	Label vessels and controls to enhance understanding

Minimizing entails reducing the hazards by using smaller quantities of hazardous substances in the reactors, distillation columns, storage vessels, and pipelines. When possible, hazardous materials should be produced and consumed in situ. This minimizes the storage and transportation of hazardous raw materials and intermediates.

Vapor released from spills can be minimized by designing dikes so that flammable and toxic materials will not accumulate around leaking tanks. Smaller tanks also reduce the hazards of a release.

While minimization possibilities are being investigated, substitutions should also be considered as an alternative or companion concept; that is, safer materials should be used in place of hazardous ones. This can be accomplished by using alternative chemistry that allows the use of less hazardous materials or less severe processing conditions. When possible, toxic or flammable solvents should be replaced with less hazardous solvents (for example, water-based paints and adhesives and aqueous or dry flowable formulations for agricultural chemicals).

Another alternative to substitution is moderation, that is, using a hazardous material under less hazardous conditions. Less hazardous conditions or less hazardous forms of a material include (1) diluting to a lower vapor pressure to reduce the release concentration, (2) refrigerating to lower the vapor pressure, (3) handling larger particle size solids to minimize dust, and (4) processing under less severe temperature or pressure conditions.

Containment buildings are sometimes used to moderate the impact of a spill of an especially toxic material. When containment is used, special precautions are included to ensure worker protection, such as remote controls, continuous monitoring, and restricted access.

Simpler plants are friendlier than complex plants because they provide fewer opportunities for error and because they contain less equipment that can cause problems. Often, the reason for complexity in a plant is the need to add equipment and automation to control the hazards. Simplification reduces the opportunities for errors and misoperation. For example, (1) piping systems can be designed to minimize leaks or failures, (2) transfer systems can be designed to minimize the potential for leaks, (3) process steps and units can be separated to prevent the domino effect, (4) fail-safe valves can be added, (5) equipment and controls can be placed in a logical order, and (6) the status of the process can be made visible and clear at all times.

The design of an inherently safe and simple piping system includes minimizing the use of sight glasses, flexible connectors, and bellows, using welded pipes for flammable and toxic chemicals and avoiding the use of threaded pipe, using spiral wound gaskets and flexible graphitetype gaskets that are less prone to catastrophic failures, and using proper support of lines to minimize stress and subsequent failures.

1-8 Seven Significant Disasters

The study of case histories provides valuable information to chemical engineers involved with safety. This information is used to improve procedures to prevent similar accidents in the future.

The seven most cited accidents (Flixborough, England; Bhopal, India; Seveso, Italy; Pasadena, Texas; Texas City, Texas; Jacksonville, Florida; and Port Wentworth, Georgia) are presented here. All these accidents had a significant impact on public perceptions and the chemical engineering profession that added new emphasis and standards in the practice of safety. Chapter 14 presents case histories in considerably more detail.

The Flixborough accident is perhaps the most documented chemical plant disaster. The British government insisted on an extensive investigation.

FLIXBOROUGH, ENGLAND

The accident at Flixborough, England, occurred on a Saturday in June 1974. Although it was not reported to any great extent in the United States, it had a major impact on chemical engineering in the United Kingdom. As a result of the accident, safety achieved a much higher priority in that country.

The Flixborough Works of Nypro Limited was designed to produce 70,000 tons per year of caprolactam, a basic raw material for the production of nylon. The process uses cyclohexane, which has properties similar to gasoline. Under the process conditions in use at Flixborough (155°C and 7.9 atm), the cyclohexane volatilizes immediately when depressurized to atmospheric conditions.

The process where the accident occurred consisted of six reactors in series. In these reactors cyclohexane was oxidized to cyclohexanone and then to cyclohexanol using injected air in the presence of a catalyst. The liquid reaction mass was gravity-fed through the series of reactors. Each reactor normally contained about 20 tons of cyclohexane.

Several months before the accident occurred, reactor 5 in the series was found to be leaking. Inspection showed a vertical crack in its stainless steel structure. The decision was made to remove the reactor for repairs. An additional decision was made to continue operating by connecting reactor 4 directly to reactor

6 in the series. The loss of the reactor would reduce the yield but would enable continued production because unreacted cyclohexane is separated and recycled at a later stage.

The feed pipes connecting the reactors were 28 inches in diameter. Because only 20-inch pipe stock was available at the plant, the connections to reactor 4 and reactor 6 were made using flexible bellows-type piping,. It is hypothesized that the bypass pipe section ruptured because of inadequate support and overflexing of the pipe section as a result of internal reactor pressures. Upon rupture of the bypass, an estimated 30 tons of cyclohexane volatilized and formed a large vapor cloud. The cloud was ignited by an unknown source an estimated 45 seconds after the release.

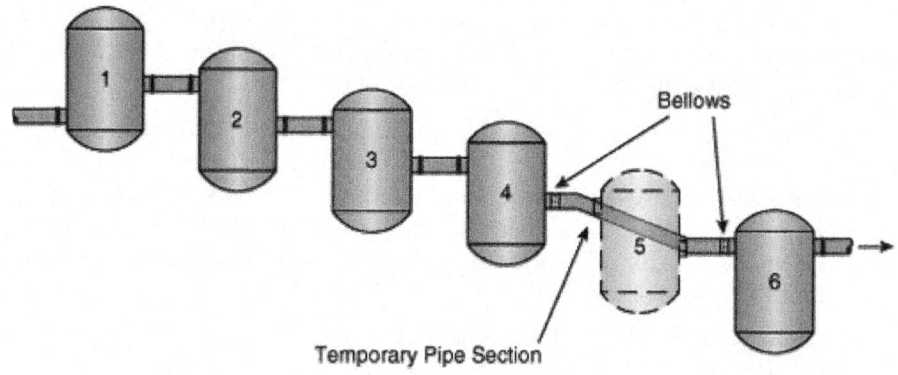

A failure of a temporary pipe section replacing reactor 5 caused
the Flixborough accident.

The resulting explosion leveled the entire plant facility, including the administrative offices. Twenty-eight people died, and 36 others were injured. Eighteen of these fatalities occurred in the main control room when the ceiling collapsed. Loss of life would have been substantially greater had the accident occurred on a weekday when the administrative offices were filled with employees. Damage extended to 1821 nearby houses and 167 shops and factories. Fifty-three civilians were reported injured. The resulting fire in the plant burned for over 10 days.

This accident could have been prevented by following proper safety procedures. First, the bypass line was installed without a safety review or adequate supervision by experienced engineering personnel. The bypass was sketched on the floor of the machine shop using chalk!

Second, the plant site contained excessively large inventories of dangerous compounds. This included 330,000 gallons of cyclohexane, 66,000 gallons of naphtha, 11,000 gallons of toluene, 26,400 gallons of benzene, and 450 gallons of gasoline. These inventories contributed to the fires after the initial blast. Finally, the bypass modification was substandard in design. As a rule, any modifications should be of the same quality as the construction of the remainder of the plant.

BHOPAL, INDIA

The Bhopal, India, accident, on December 3, 1984, has received considerably more attention than the Flixborough accident. This is due to the more than 2000 civilian casualties that resulted.

The Bhopal plant is in the state of Madhya Pradesh in central India. The plant was partially owned by Union Carbide and partially owned locally.

The nearest civilian inhabitants were 1.5 miles away when the plant was constructed. Because the plant was the dominant source of employment in the area, a shantytown eventually grew around the immediate area.

The plant produced pesticides. An intermediate compound in this process is methyl isocyanate (MIC). MIC is an extremely dangerous compound. It is reactive, toxic, volatile, and flammable. The maximum exposure concentration of MIC for workers over an 8-hour period is 0.02 ppm (parts per million). Individuals exposed to concentrations of MIC vapors above 21 ppm experience severe irritation of the nose and throat. Death at large concentrations of vapor is due to respiratory distress.

MIC demonstrates a number of dangerous physical properties. Its boiling point at atmospheric conditions is 39.1°C, and it has a vapor pressure of 348 mm Hg at 20°C. The vapor is about twice as heavy as air, ensuring that the vapors will stay close to the ground once released.

MIC reacts exothermically with water. Although the reaction rate is slow, with inadequate cooling the temperature will increase and the MIC will boil. MIC storage tanks are typically refrigerated to prevent this problem.

The unit using the MIC was not operating because of a local labor dispute. Somehow a storage tank containing a large amount of MIC became contaminated with water or some other substance. A chemical reaction heated the MIC to a temperature past its boiling point. The MIC vapors traveled through a pressure relief system and into a scrubber and flare system installed to consume the MIC in the event of a release. Unfortunately, the scrubber and flare systems were not operating, for a variety of reasons. An estimated 25 tons of toxic MIC vapor was released. The toxic cloud spread to the adjacent town, killing over 2000 civilians and injuring an estimated 20,000 more. No plant workers were injured or killed. No plant equipment was damaged.

The exact cause of the contamination of the MIC is not known. If the accident was caused by a problem with the process, a well-executed safety review could have identified the problem. The scrubber and flare system should have been fully operational to prevent the release. Inventories of dangerous chemicals, particularly intermediates, should also have been minimized.

The reaction scheme used at Bhopal is shown at the top of and includes the dangerous intermediate MIC. An alternative reaction scheme is shown at the bottom of the figure and involves a less dangerous chloroformate intermediate. Another solution is to redesign the process to reduce the inventory of hazardous

MIC. One such design produces and consumes the MIC in a highly localized area of the process, with an inventory of MIC of less than 20 pounds.

Methyl isocyanate route

$$CH_3NH_2 + COCl_2 \longrightarrow CH_3N = C = O + 2HCl$$

Methylamine   Phosgene   Methyl isocyanate

α-Naphthol   Carbaryl

Nonmethyl isocyanate route

α-Naphthol chloroformate

The upper reaction is the methyl isocyanate route used at Bhopal. The lower reaction suggests an alternative reaction scheme using a less hazardous intermediate. Adapted from (Feb. 11, 1985), p. 30.

Seveso, Italy

Seveso is a small town of approximately 17,000 inhabitants, 15 miles from Milan, Italy. The plant was owned by the Icmesa Chemical Company. The product was hexachlorophene, a bactericide, with trichlorophenol produced as an intermediate. During normal operation, a small amount of TCDD (2,3,7,8-tetrachlorodibenzoparadioxin) is produced in the reactor as an undesirable side-product.

TCDD is perhaps the most potent toxin known to humans. Animal studies have shown TCDD to be fatal in doses as small as 10^{-9} times the body weight.

Because TCDD is also insoluble in water, decontamination is difficult. Nonlethal doses of TCDD result in chloracne, an acne-like disease that can persist for several years.

On July 10, 1976, the trichlorophenol reactor went out of control, resulting in a higher than normal operating temperature and increased production of TCDD. An estimated 2 kg of TCDD was released through a relief system in a white cloud over Seveso. A subsequent heavy rain washed the TCDD into the soil. Approximately 10 square miles were contaminated.

Because of poor communications with local authorities, civilian evacuation was not started until several days later. By then, over 250 cases of chloracne were reported. Over 600 people were evacuated, and an additional 2000 people were given blood tests. The most severely contaminated area immediately adjacent to the plant was fenced, the condition it remains in today.

TCDD is so toxic and persistent that for a smaller but similar release of TCDD in Duphar, India, in 1963 the plant was finally disassembled brick by brick, encased in concrete, and dumped into the ocean. Less than 200 g of TCDD was released, and the contamination was confined to the plant. Of the 50 men assigned to clean up the release, 4 eventually died from the exposure.

The Seveso and Duphar accidents could have been avoided if proper containment systems had been used to contain the reactor releases. The proper application of fundamental engineering safety principles would have prevented the two accidents. First, by following proper procedures, the initiation steps would not have occurred. Second, by using proper hazard evaluation procedures, the hazards could have been identified and corrected before the accidents occurred.

PASADENA, TEXAS

A massive explosion in Pasadena, Texas, on October 23, 1989, resulted in 23 fatalities, 314 injuries, and capital losses of over $715 million. This explosion occurred in a high-density polyethylene plant after the accidental release of 85,000 pounds of a flammable mixture containing ethylene, isobutane, hexane, and hydrogen. The release formed a large gas cloud instantaneously because the system was under high pressure and temperature. The cloud was ignited about 2 minutes after the release by an unidentified ignition source.

The damage resulting from the explosion made it impossible to reconstruct the actual accident scenario. However, evidence showed that the standard operating procedures were not appropriately followed.

The release occurred in the polyethylene product takeoff system. Usually the polyethylene particles (product) settle in the settling leg and are removed through the product takeoff valve. Occasionally, the product plugs the settling leg, and the plug is removed by maintenance personnel. The normal—and safe—procedure includes closing the DEMCO valve, removing the air lines, and locking the valve in the closed position. Then the product takeoff valve is removed to give access to the plugged leg.

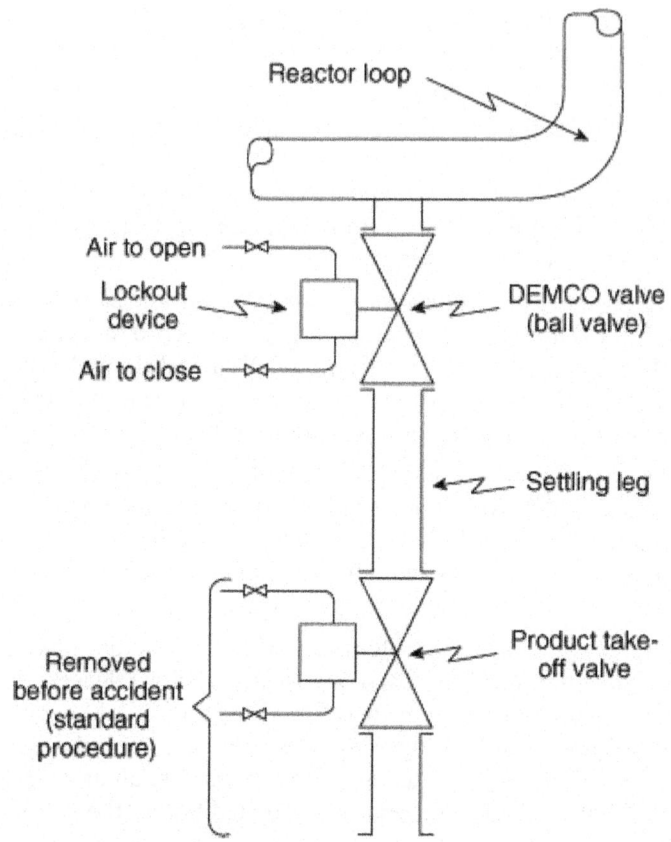

Reactor loop

Air to open

Lockout device

Air to close

DEMCO valve (ball valve)

Settling leg

Removed before accident (standard procedure)

Product take-off valve

Polyethylene plant settling leg and product takeoff system.

The accident investigation evidence showed that this safe procedure was not followed; specifically, the product takeoff valve was removed, the DEMCO valve was in the open position, and the lockout device was removed. This scenario was a serious violation of well-established and well-understood procedures and created the conditions that permitted the release and subsequent explosion.

The OSHA investigation found that (1) no process hazard analysis had been performed in the polyethylene plant, and as a result, many serious safety deficiencies were ignored or overlooked; (2) the single-block (DEMCO) valve on the settling leg was not designed to fail to a safe closed position when the air failed; (3) rather than relying on a single-block valve, a double block and bleed valving arrangement or a blind flange after the single-block valve should have been used; (4) no provision was made for the development, implementation, and enforcement of effective permit systems (for example, line opening); and (5) no permanent combustible gas detection and alarm system was located in the region of the reactors.

Other factors that contributed to the severity of this disaster were also cited: (1) proximity of high-occupancy structures (control rooms) to hazardous operation, (2) inadequate separation between buildings, and (3) crowded process equipment.

Texas City, Texas

A petroleum refinery had large explosions on March 23, 2005, that killed 15 workers and injured about 180. The explosions were the result of a sudden release of flammable liquid and vapor from an open vent stack in the refinery's isomerization (ISOM) unit. The ISOM unit converts pentane and hexane into isopentane and isohexane (gasoline additive). The unit works by heating the pentane and hexane in the presence of a catalyst. This unit includes a splitter tower and associated process equipment, which is used to prepare the hydrocarbon feed of the isomerization reactor.

This accident was during the startup of this ISOM process unit. In this startup, hydrocarbons were pumped into the splitter tower for three hours without any liquid being removed and transferred to storage (which should have happened). As a result, the 164-foot-tall tower was overfilled. The resulting high pressure activated three pressure relief valves, and the liquid was discharged to a vented blowdown drum. The blowdown drum overfilled with hydrocarbons, producing a geyser-like release from the vented stack. The flammable hydrocarbons pooled on the ground, releasing vapors that ignited, resulting in multiple explosions and fires. Many of those killed were working in or around two contractor office trailers located near a blowdown drum.

The CSB investigation identified the following major findings: (1) the occupied trailers were sited in an unsafe location (all 15 fatalities occurred in or around two contractor trailers); (2) the ISOM unit should not have been started up because there were existing and known problems that should have been repaired before a startup (known equipment malfunctions included a level indicator and alarm, and a control valve); and (3) previously there were at least four other serious releases of flammables out of this blowdown drum vent, and even though these serious near-misses revealed the existing hazard, no effective investigations were conducted nor were appropriate design changes made (a properly designed flare system would have burned these effluents to prevent this unsafe release of the flammable liquid and combustible vapors).

Jacksonville, Florida

CSB investigated an accident that occurred in a chemical manufacturing plant (gasoline additive) on December 19, 2007. A powerful explosion and fire killed 4 employees and injured 32, including 4 employees and 28 members of the public who were working in surrounding businesses. This plant blended and sold printing solvents and started to manufacture methylcyclopentadienyl manganese tricarbonyl (MCMT) in a 2500-gallon batch reactor in January of 2004.

The accident occurred while the plant was producing its 175th batch of MCMT. The process included two exothermic reactions, the first a necessary step in the production of MCMT, and the second an unwanted side reaction that occurs at about 390°F, which is slightly higher than the normal operating temperature. The reactor cooling failed (line blockage or valve failure), and the temperature increased, setting off both runaway reactions uncontrollably. About ten minutes after the initial cooling failure, the reactor burst and its contents exploded due to the uncontrolled high temperatures and pressures. The pressure burst the reactor and the reactor's contents exploded with a TNT equivalent to 1400 pounds of TNT. Debris from the reactor was found up to one mile away, and the explosion damaged buildings within one-quarter mile of the facility.

CSB found that (1) the cooling system was susceptible to only single-point failures due to the lack of design redundancy, (2) the reactor relief system was incapable of relieving the pressure from the runaway reactions, and (3) despite a number of previous and similar near-misses the company employees failed to recognize the hazards of the runaway reactions associated with this manufacturing process (even though the two owners of the company had undergraduate degrees in chemistry and chemical engineering).

The CSB recommendations in this accident investigation report focused on improving the education of chemical engineering students on the hazards of reactive chemicals.

Port Wentworth, Georgia

On February 7, 2008, a series of sugar dust explosions at a sugar manufacturing facility resulted in 14 fatalities and 36 injuries.. This refinery converted raw sugarcane into granulated sugar. A system of screw and belt conveyors and bucket elevators transported granulated sugar from the refinery to storage silos, and to specialty sugar processing areas.

A recently installed steel cover panel on the belt conveyor allowed explosive concentrations of sugar dust to accumulate inside the enclosure. The first dust explosion occurred in this enclosed steel belt conveyor located below the sugar silos. An overheated bearing in the steel belt conveyor was the most likely ignition source. This primary explosion dispersed sugar dust that had accumulated on the floors and elevator horizontal surfaces, propagating more explosions throughout the buildings. Secondary dust explosions occurred throughout the packing buildings, parts of the refinery, and the loading buildings. The pressure waves from the explosions heaved thick concrete floors and collapsed brick walls, blocking stairwell and other exit routes.

The CSB investigation identified three major causes: (1) The conveying equipment was not designed to minimize the release of sugar dust and eliminate all ignition sources in the work areas; (2) housekeeping practices were poor; and (3) the company failed to correct the ongoing and known hazardous conditions, despite the well-known and broadly published hazards associated with combustible dusts.

Prior to this Port Wentworth accident, CSB undertook a study in 2005 concerning the extent of the industrial dust explosion problem. They identified 200 fires and explosions due to dusts over a 25-year period that took 100 lives and caused 600 injuries. The tragic event in Port Wentworth demonstrates that dust explosions in industry continue to be a problem.

Problems

1-1. An employee works in a plant with a FAR of 4. If this employee works a 4-hr shift, 200 days per year, what is the expected deaths per person per year?

1-2. Three process units are in a plant. The units have FARs of 0.5, 0.3, and 1.0, respectively.

    a. What is the overall FAR for the plant, assuming worker exposure to all three units simultaneously?

    b. Assume now that the units are far enough apart that an accident in one would not affect the workers in another. If a worker spends 20% of his time in process area 1, 40% in process area 2, and 40% in process area 3, what is his overall FAR?

1-3. Assuming that a car travels at an average speed of 50 miles per hour, how many miles must be driven before a fatality is expected?

1-4. A worker is told her chances of being killed by a particular process are 1 in every 500 years. Should the worker be satisfied or alarmed? What is the FAR (assuming normal working hours) and the deaths per person per year? What should her chances be, assuming an average chemical plant?

1-5. A plant employs 1500 full-time workers in a process with a FAR of 5. How many industrial-related deaths are expected each year?

1-6. Consider Example 1-4. How many hours must be traveled by car for each hour of rock climbing to make the risk of fatality by car equal to the risk of fatality by rock climbing?

1-7. Identify the initiation, propagation, and termination steps for the following accident reports. Suggest ways to prevent and contain the accidents.

    a. A contractor accidentally cut into a 10-in propane line operating at 800 psi at a natural gas liquids terminal. The large vapor cloud estimated to cover an area of 44 acres was ignited about 4–5 min later by an unknown source. Liquid products from 5 of 26 salt dome caverns fed the fire with an estimated 18,000–30,000 gal of LPGs for almost 6 hr before being blocked in and the fires extinguished. Both engine-driven fire pumps failed, one because intense radiated heat damaged its ignition wires and the other because the explosion broke a sight glass fuel gauge, spilling diesel fuel, which ignited, destroying the fire pump engine.

    b. An alkylation unit was being started up after shutdown because of an electrical outage. When adequate circulation could not be maintained in a deisobutanizer heater circuit, it was decided to clean the strainer. Workers had depressurized the pipe and removed all but three of the flange bolts when a pressure release blew a black material from the flange, followed by butane vapors. These vapors were carried to a furnace 100 ft away, where they ignited, flashing back to the flange. The ensuing fire exposed a fractionation tower and horizontal receiver drums. These drums exploded, rupturing pipelines, which added more fuel. The explosions and heat caused loss of insulation from the 8-ft x 122-ft fractionator tower, causing it to weaken and fall across two major pipelines, breaking piping — which added more fuel to the fire. Extinguishment, achieved basically by isolating the fuel sources, took 2 1/2; hours.

The fault was traced to a 10-in valve that had been prevented from closing the last 3/4;-inch by a fine powder of carbon and iron oxide. When the flange was opened, this powder blew out, allowing liquid butane to be released.

1-8. The airline industry claims commercial airline transport has fewer deaths per mile than any other means of transportation. Do the accident statistics support this claim? In 1984 the airline industry posted 4 deaths per 10,000,000 passenger miles. What additional information is required to compute a FAR? a fatality rate?

1-9. A university has 1200 full-time employees. In a particular year this university had 38 reportable lost-time injuries with a resulting 274 lost workdays. Compute the OSHA incidence rate based on injuries and lost workdays.

1. An explosion has occurred in your plant and an employee has been killed. An investigation shows that the accident was the fault of the dead employee, who manually charged the wrong ingredient to a reactor vessel. What is the appropriate response from the following groups?

    a.   The other employees who work in the process area affected.

    b.   The other employees elsewhere in the plant site.

    c.   Middle management.

    d.   Upper management.

    e.   The president of the company.

    f.   The union.

1-1. You have just begun work at a chemical plant. After several weeks on the job you determine that the plant manager runs the plant with an iron fist. He is a few years away from retirement after working his way up from the very bottom. Also, a number of unsafe practices are performed at the plant, including some that could lead to catastrophic results. You bring up these problems to your immediate supervisor, but he decides to do nothing for fear that the plant manager will be upset. After all, he says, "We've operated this plant for 40 years without an accident." What would you do in this situation?

1-1. a. You walk into a store and after a short while you decide to leave, preferring not to do any business there. What did you observe to make you leave? What conclusions might you reach about the attitudes of the people who manage and operate this store?

b. You walk into a chemical plant and after a short while you decide to leave, fearing that the plant might explode at any moment. What did you observe to make you leave? What conclusions might you reach about the attitudes of the people who manage and operate this chemical plant?

Comment on the similarities of parts a and b.

1-1. A large storage tank is filled manually by an operator. The operator first opens a valve on a supply line and carefully watches the level on a level indicator until the tank is filled (a long time later). Once the filling is complete, the operator closes the valve to stop the filling. Once a year the operator is distracted and the tank is overfilled. To prevent this, an alarm was installed on the level gauge to alert the operator to a high-level condition. With the installation of the alarm, the tank now overfills twice per year. Can you explain?

1-1. Careful numbering of process equipment is important to avoid confusion. On one unit the equipment was numbered J1001 upward. When the original allocation of numbers ran out the new equipment was numbered JA1001 upward. An operator was verbally told to prepare pump JA1001 for repairs. Unfortunately, he prepared pump J1001 instead, causing an upset in the plant. What happened?

1-2. A cover plate on a pump housing is held in place by eight bolts. A pipefitter is instructed to repair the pump. The fitter removes all eight bolts only to find the cover plate stuck on the housing. A screwdriver is used to pry off the cover. The cover flies off suddenly, and toxic liquid sprays throughout the work area. Clearly the pump unit should have been isolated, drained, and cleaned before repair. There is, however, a better procedure for removing the cover plate. What is this procedure?

1-21. The liquid level in a tank 10 m in height is determined by measuring the pressure at the bottom of the tank. The level gauge was calibrated to work with a liquid having a specific gravity of 0.9. If the usual liquid is replaced with a new liquid with a specific gravity of 0.8, will the tank be overfilled or underfilled? If the actual liquid level is 8 m, what is the reading on the level gauge? Is it possible that the tank will overflow without the level gauge indicating the situation?

1-22. One of the categories of inherent safety is simplification/error tolerance. What instrumentation could you add to the tank described in Problem 1-21 to eliminate problems?

1-23. Pumps can be shut-in by closing the valves on the inlet and outlet sides of the pump. This can lead to pump damage and/or a rapid increase in the temperature of the liquid shut inside the pump. A particular pump contains 4 kg of water. If the pump is rated at 1 HP, what is the maximum temperature increase expected in the water in °C/hr? Assume a constant water heat capacity of 1 kcal/kg/°C. What will happen if the pump continues to operate?

1-24. Water will flash into vapor almost explosively if heated under certain conditions.

    a. What is the ratio in volume between water vapor at 300 K and liquid water at 300 K at saturated conditions?

    b. Hot oil is accidentally pumped into a storage vessel. Unfortunately, the tank contains residual water, which flashes into vapor and ruptures the tank. If the tank is 10 m in diameter and 5 m high, how many kilograms of water at 300 K are required to produce enough water vapor to pressurize the tank to 8 in of water gauge pressure, the burst pressure of the tank?

1-25. Another way of measuring accident performance is by the LTIR, or lost-time injury rate. This is identical to the OSHA incidence rate based on incidents in which the employee is unable to continue their normal duties. A plant site has 1200 full-time employees working 40 hr/week and 50 weeks/yr. If the plant had 2 lost-time incidents last year, what is the LTIR?

1-26. A car leaves New York City and travels the 2800-mi distance to Los Angeles at an average speed of 50 mph. An alternative travel plan is to fly on a commercial airline for 4 1/2;hr. What are the FARs for the two methods of transportation? Which travel method is safer, based on the FAR?

1-27. A column was used to strip low-volatile materials from a high-temperature heat transfer fluid. During a maintenance procedure, water was trapped between two valves. During normal operation, one valve was opened and the hot oil came in contact with the cold water. The result was almost sudden vaporization of the water, followed by considerable damage to the column. Consider liquid water at 25°C and 1 atm. How many times does the volume increase if the water is vaporized at 100°C and 1 atm?

1-28. Large storage tanks are designed to withstand low pressures and vacuums. Typically they are constructed to withstand no more than 8 in of water gauge pressure and 2.5 in of water gauge vacuum. A particular tank is 30 ft in diameter.

   a. If a 200-lb person stands in the middle of the tank roof, what is the resulting pressure (in inches of water gauge) if the person's weight is distributed across the entire roof?

   b. If the roof was flooded with 8 in of water (equivalent to the maximum pressure), what is the total weight (in pounds) of the water?

   c. A large storage tank was sucked in when the vent to the outside became plugged and the operator turned on the pump to empty the tank. How did this happen?

Note: A person can easily blow to a pressure of greater than 20 in of water gauge.

1-29. A 50-gal drum with bulged ends is found in the storage yard of your plant. You are unable to identify the contents of the drum. Develop a procedure to handle this hazard. There are many ways to solve this problem. Please describe just one approach.

1-30. The plant has been down for extensive maintenance and repair. You are in charge of bringing the plant up and on-line. There is considerable pressure from the sales department to deliver product. At about 4 *A.M.* a problem develops. A slip plate or blind has accidentally been left in one of the process lines. An experienced maintenance person suggests that she can remove the slip plate without depressurizing the line. She said that she routinely performed this operation years ago. Since you are in charge, what would you do?

1-31. Gasoline tank trucks are load restricted in that the tank must never be between 20% and 80% full when traveling. Or it must be below 20% and above 80%. Why?

1-32. In 1891 the copper industry in Michigan employed 7702 workers. In that year there were 28 fatalities in the mines. Estimate the FAR for this year, assuming that the workers worked 40-hour weeks and 50 weeks per year. Compare the result to the published FAR for the chemical industry.

1-33. The Weather Channel reports that, on average, about 42 Americans are killed by lightning each year. The current population of the U.S. is about 300 million people. Which accident index is suitable for this information: FAR, OSHA incident rate, or deaths per person per year? Why? Calculate the value of the selected index and compare it to published values.

1-34. The CSB video "Preventing Harm from Sodium Hydrosulfide" presents an incident involving sodium hydrosulfide (NaSH) and hydrogen sulfide (H_2S). Go on-line and find at least two material safety data sheets (MSDS) for both of these chemicals. Tabulate the following physical properties for these chemicals at room temperature and pressure, if available: physical state density, PEL, TLV, and vapor pressure. List any other concerns that might be apparent from the MSDS. Which of these properties are of major concern in using these chemicals?

Chapter 2

FUNDAMENTAL OF TOXICOLOGY IN CHEMICAL PROCESS

Toxicology (from the Ancient Greek words *toxico* "poisonous" and *logos*) is a branch of biology, chemistry, and medicine (more specifically pharmacology) concerned with the study of the adverse effects of chemicals on living organisms. [1] A **toxicologist** is a scientist or medical personal who specializes in the study of symptoms, mechanisms, treatments and detection of venoms and toxins; especially the poisoning of people.

It also studies the harmful effects of chemical, biological and physical agents in biological systems that establishes the extent of damage in living organisms.

HISTORY

Dioscorides, a Greek physician in the court of the Roman emperor Nero, made the first attempt to classify plants according to their toxic and therapeutic effect. Ibn Wahshiya wrote the *Book on Poisons* in the 9th or 10th century.

Mathieu Orfila is considered the modern father of toxicology, having given the subject its first formal treatment in 1813 in his *Traité des poisons*, also called *Toxicologie générale*.

In 1850, Jean Stas gave the evidence that the Belgian Count Hippolyte Visart de Bocarmé killed his brother-in-law by poisoning him with nicotine.

Theophrastus Phillipus Auroleus Bombastus von Hohenheim (1493–1541) (also referred to as Paracelsus, from his belief that his studies were above or beyond the work of Celsus – a Roman physician from the first century) is also considered "the father" of toxicology.[6] He is credited with the classic toxicology maxim, "*Alle Dinge sind Gift und nichts ist ohne Gift; allein die Dosis macht, dass ein Ding kein Gift ist.*" which translates as, "All things are poison and nothing is without poison; only the dose makes a thing not a poison." This is often condensed to: "The dose makes the poison" or in Latin "Sola dosis facit venenum".

Basic Traditional Toxicology

The relationship between dose and its effects on the exposed organism is of high significance in toxicology. The chief criterion regarding the toxicity of a chemical is the dose, *i.e.* the amount of exposure to the substance. All substances are toxic under the right conditions. The term LD_{50} refers to the dose of a toxic substance that kills 50 percent of a test population (typically rats or other surrogates when the test concerns human toxicity).

The conventional relationship (more exposure equals higher risk) has been challenged in the study of endocrine disruptors. Toxicity is species-specific, lending cross-species analysis problematic. Newer methods are available to bypass animal-testing.

A nontechnical popularization of traditional toxicology is available in the book *The Dose Makes the Poison*.

Factors that influence chemical toxicity:

- Dosage
- Both large single exposures (acute) and continuous small exposures (chronic) are studied.
- Route of exposure
- Ingestion, inhalation or skin absorption
- Other factors
- Species
- Age
- Sex
- Health
- Environment
- Individual characteristics

Foods safe for humans are not necessarily safe for pets. A young healthy pregnant woman in a supportive environment has a different set of chemical sensitivities than an aged homeless male drug addict. Chemicals safe to drink may not be safe to inject. Eating a peanut is life-threatening for some.

The classic experimental tool of toxicology is animal testing. Alternative tests have been and are being developed. One example of a test that has been developed is Corrositex, a toxicology test that is much more accurate than animal testing. Separate test protocols are used for acute and chronic toxicity, irritation, sensitization (allergies), reproductive toxicity and carcinogenesis (cancer).

Few antidotes to poisons exist. Treatment usually consists of removing the poison, repairing damage and providing life support.

The testing of one chemical for its cancer-causing properties took 5 years, cost more than $6.5 million in 1980 and utilized 24,000 mice.

Dose Response Complexities

Most chemicals display a classic dose response curve – at a low dose (below a threshold), no effect is observed. Some show a phenomenon known as sufficient challenge – a small exposure produces animals that "grow more rapidly, have better general appearance and coat quality, have fewer tumors, and live longer than the control animals". A few chemicals have no well-defined safe level of exposure. These are treated with special care. Some chemicals are subject to bioaccumulation as they are stored in rather than being excreted from the body; these also receive special consideration.

Toxicologist Duties

Toxicologists do many different duties including research in the academic, nonprofit and industrial fields, product safety evaluation, consulting, public service and legal regulation. In and order to research and assess the effects of chemicals, toxicologists perform carefully designed studies and experiments. These experiments help identify the specific amount of a chemical that may cause harm and potential risks of being near or using products that contain certain chemicals. Research projects may range from assessing the effects of toxic pollutants on the environment to evaluating how the human immune system responds to chemical compounds within pharmaceutical drugs.

While the basic duties of toxicologists are to determine the effects of chemicals on organisms and their surroundings, specific job duties may vary based on industry and employment. For example, forensic toxicologists may look for toxic substances in a crime scene, whereas aquatic toxicologists may analyze the toxicity level of wastewater.

Toxicology Starting Pay

The salary for jobs in toxicology is dependent on several factors, including level of schooling, specialization, experience. The U.S. Bureau of Labor Statistics (BLS) notes that jobs for biological scientists, which generally include toxicologists, were expected to increase by 21% between 2008 and 2018. The BLS notes that this increase could be due to research and development growth in biotechnology, as well as budget increases for basic and medical research in biological science. Salary.com notes that the median annual salary for toxicologists as of 2011 was $84,000.

Requirements for a Toxicologist

To work as a toxicologist one should obtain a degree in toxicology or a related degree like biology, chemistry or biochemistry. Bachelor's degree programs in toxicology cover the chemical makeup of toxins and their effects on biochemistry, physiology and ecology. After introductory life science courses are complete, students typically enroll in labs and apply toxicology principles to research and other studies. Advanced students delve into specific sectors, like the pharmaceutical industry or law enforcement, which apply methods of toxicology in their work.

The Society of Toxicology (SOT) recommends that undergraduates in post-secondary schools that don't offer a bachelor's degree in toxicology consider attaining a degree in biology or chemistry. Additionally, the SOT advises aspiring toxicologists to take statistics and mathematics courses, as well as gain laboratory experience through lab courses, student research projects and internships.

Toxicology is the study of the adverse effects of chemicals on living organisms. It is the study of symptoms, mechanisms, treatments and detection of poisoning, especially the poisoning of people.

The relationship between dose and its effects on the exposed organism is of high significance in toxicology. The chief criterion regarding the toxicity of a chemical is the dose, *i.e.* the amount of exposure to the substance.

All substances are toxic under the right conditions. The term LD_{50} refers to the dose of a toxic substance that kills 50 percent of a test population (typically rats or other surrogates when the test concerns human toxicity).

LD_{50} estimations in animals are no longer required for regulatory submissions as a part of pre-clinical development package.

The conventional relationship (more exposure equals higher risk) has been challenged in the study of endocrine disruptors.

There are various specialized subdisciplines within the field of toxicology that concern diverse chemical and biological aspects of this area.

For example, toxicogenomics involves applying molecular profiling approaches to the study of toxicology.

Other areas include Aquatic toxicology, Chemical toxicology, Ecotoxicology, Environmental toxicology, Forensic toxicology, and Medical toxicology.

Chemical toxicology is a scientific discipline involving the study of structure and mechanism related to the toxic effects of chemical agents, and encompasses technology advances in research related to chemical aspects of toxicology.

Research in this area is strongly multidisciplinary, spanning computational chemistry and synthetic chemistry, proteomics and metabolomics, drug discovery, drug metabolism and mechanisms of action, bioinformatics, bioanalytical chemistry, chemical biology, and molecular epidemiology.

Toxicology can be defined as that branch of science that deals with poisons, and a poison can be defined as any substance that causes a harmful effect when administered, either by accident or design, to a living organism. By convention, toxicology also includes the study of harmful effects caused by physical phenomena, such as radiation of various kinds and noise. In practice, however, many complications exist beyond these simple definitions, both in bringing more precise meaning to what constitutes a poison and to the measurement of toxic effects. Broader definitions of toxicology, such as "the study of the detection, occurrence, properties, effects, and regulation of toxic substances," although more descriptive, do not resolve the difficulties. Toxicity itself can rarely, if ever, be defined as a single molecular event but is, rather, a cascade of events starting with exposure, proceeding through distribution and metabolism, and ending with interaction with cellular macromolecules (usually DNA or protein) and the expression of a toxic end point. This sequence may be mitigated by excretion and repair. It is to the complications, and to the science behind them and their resolution, that this textbook is dedicated, particularly to the how and why certain substances cause disruptions in biologic systems that result in toxic effects. Taken together, these difficulties and their resolution circumscribe the perimeter of the science of toxicology.

The study of toxicology serves society in many ways, not only to protect humans and the environment from the deleterious effects of toxicants but also to facilitate the development of more selective toxicants such as anticancer and other clinical drugs and pesticides. Poison is a quantitative concept, almost any substance being harmful at some doses but, at the same time, being without harmful effect at some lower dose. Between these two limits there is a range of possible effects, from subtle long-term chronic toxicity to immediate lethality. Vinyl chloride may be taken as an example. It is a potent hepatotoxicant at high doses, a carcinogen with a long latent period at lower doses, and apparently without effect at very low doses. Clinical drugs are even more poignant examples because, although therapeutic and highly beneficial at some doses, they are not without deleterious side effects and may be lethal at higher doses. Aspirin (acetylsalicylic acid), for example, is a relatively safe drug at recommended doses and is taken by millions of people worldwide. At the same time, chronic use can cause deleterious effects on the gastric mucosa, and it is fatal at a dose of about 0.2 to 0.5 g/kg. Approximately 15% of reported accidental deaths from poisoning in children result from ingestion of salicylates, particularly aspirin.

The importance of dose is well illustrated by metals that are essential in the diet but are toxic at higher doses. Thus iron, copper, magnesium, cobalt, manganese, and zinc can be present in the diet at too low a level (deficiency), at an appropriate level (maintenance), or at too high a level (toxic). The question of dose-response relationships is fundamental to toxicology. The definition of a poison, or toxicant, also involves a qualitative biological aspect because a compound, toxic to one species or genetic strain, may be relatively harmless to another. For example, carbon tetrachloride, a potent hepatotoxicant in many species, is relatively harmless to the chicken. Certain strains of rabbit can eat Belladonna with impunity while others cannot. Compounds may be toxic under some circumstances but not others or, perhaps, toxic in combination with another compound but nontoxic alone. The methylenedioxyphenyl insecticide synergists, such as piperonyl butoxide, are of low toxicity to both insects and mammals when administered alone but are, by virtue of their ability to inhibit xenobiotic-metabolizing enzymes, capable of causing dramatic increases in the toxicity of other compounds.

Modes of Toxic Action

This includes the consideration, at the fundamental level of organ, cell and molecular function, of all events leading to toxicity in vivo: uptake, distribution, metabolism, mode of action, and excretion. The term mechanism of toxic action is now more generally used to describe an important molecular event in the cascade of events leading from exposure to toxicity, such as the inhibition of acetylcholinesterase in the toxicity of organophosphorus and carbamate insecticides. Important aspects include the following:

- Biochemical and molecular toxicology consider events at the biochemical and molecular levels, including enzymes that metabolize xenobiotics, generation of reactive intermediates, interaction of xenobiotics or their metabolites.

- Behavioral toxicology deals with the effects of toxicants on animal and human behavior, which is the final integrated expression of nervous function in the intact animal. This involves both the peripheral and central nervous systems, as well as effects mediated by other organ systems, such as the endocrine glands.

- Nutritional toxicology deals with the effects of diet on the expression of toxicity and with the mechanisms of these effects.

- Carcinogenesis includes the chemical, biochemical, and molecular events that lead to the large number of effects on cell growth collectively known as cancer.

- Teratogenesis includes the chemical, biochemical, and molecular events that lead to deleterious effects on development

- . Mutagenesis is concerned with toxic effects on the genetic material and the inheritance of these effects.

- Organ toxicity considers effects at the level of organ function (neurotoxicity, hepatotoxicity, nephrotoxicity.

Measurement of Toxicants and Toxicity. These important aspects deal primarily with analytical chemistry, bioassay, and applied mathematics; they are designed to provide the methodology to answer certain critically important questions. Is the substance likely to be toxic? What is its chemical identify? How much of it is present? How can we assay its toxic effect, and what is the minimum level at which this toxic effect can be detected? A number of important fields are included:

1. Analytical toxicology is a branch of analytical chemistry concerned with the identification and assay of toxic chemicals and their metabolites in biological and environmental materials.

2. Toxicity testing involves the use of living systems to estimate toxic effects. It covers the gamut from short-term tests for genotoxicity such as the Ames test and cell culture techniques to the use of intact animals for a variety of tests from acute toxicity to lifetime chronic toxicity. Although the term "bioassay" is used properly only to describe the use of a living organism to quantitate the amount of a particular toxicant present, it is frequently used to describe any in vivo toxicity test.

3. Toxicologic pathology is the branch of pathology that deals with the effects of toxic agents manifested as changes in subcellular, cellular, tissue, or organ morphology.

5. Biomathematics and statistics relate to many areas of toxicology. They deal with data analysis, the determination of significance, and the formulation of risk estimates and predictive models.

6. Epidemiology as it applies to toxicology, is of great importance as it deals with the relationship between chemical exposure and human disease.

Chemical Use Classes. This includes the toxicology aspects of the development of new chemicals for commercial use. In some of these use classes, toxicity, at least to some organisms, is a desirable trait; in others, it is an undesirable side effect. Use classes are not composed entirely of synthetic chemicals; many natural products are isolated and used for commercial and other purposes and must be subjected to the same toxicity testing as that required for synthetic chemicals. Examples of such natural products include the insecticide, pyrethrin, the clinical drug, digitalis, and the drug of abuse, cocaine.

Toxicology is the study of poisons, or, more comprehensively, the identification and quantification of adverse outcomes associated with exposures to physical agents, chemical substances and other conditions. As such, toxicology draws upon most of the basic biological sciences, medical disciplines, epidemiology and some areas of chemistry and physics for information, research designs and methods. Toxicology ranges from basic research investigations on the mechanism of action of toxic agents through the development and interpretation of standard tests characterizing the toxic properties of agents. Toxicology provides important information for both medicine and epidemiology in understanding aetiology and in providing information as to the plausibility of observed associations between exposures, including occupations, and disease. Toxicology can be divided into standard disciplines, such as clinical, forensic, investigative and regulatory toxicology; toxicology can be considered by target organ system or process, such as immunotoxicology or genetic toxicology; toxicology can be presented in functional terms, such as research, testing and risk assessment.

It is a challenge to propose a comprehensive presentation of toxicology in this Encyclopaedia. This chapter does not present a compendium of information on toxicology or adverse effects of specific agents. This latter information is better obtained from databases that are continually updated, as described in the last section of this chapter. Moreover, the chapter does not attempt to set toxicology within specific subdisciplines, such as forensic toxicology. It is the premise of the chapter that the information provided is relevant to all types of toxicological endeavours and to the use of toxicology in various medical specialities and fields. In this chapter, topics are based primarily upon a practical orientation and integration with the intent and purpose of the Encyclopaedia as a whole. Topics are also selected for ease of cross-reference within the Encyclopaedia.

In modern society, toxicology has become an important element in environmental and occupational health. This is because many organizations, governmental and non-governmental, utilize information from toxicology to evaluate and regulate hazards in the workplace and nonoccupational environment. As part of prevention strategies, toxicology is invaluable, since it is the source of information on potential hazards in the absence of widespread human exposures. Toxicological methods are also widely used by industry in product development, to provide information useful in the design of specific molecules or product formulations.

The chapter begins with five articles on general principles of toxicology, which are important to the consideration of most topics in the field. The first general principles relate to understanding relationships between external exposure and internal dose. In modern terminology, "exposure" refers to the concentrations or amount of a substance presented to individuals or populations — amounts found in specific volumes of air or water, or in masses of soil. "Dose" refers to the concentration or amount of a substance inside an exposed person or organism. In occupational health, standards and guidelines are often set in terms of exposure, or allowable limits on concentrations in specific situations, such as in air in the workplace. These exposure limits are predicated upon assumptions or information on the relationships between exposure and dose; however, often information on internal dose is unavailable. Thus, in many studies of occupational health, associations can be drawn only between exposure and response or effect. In a few instances, standards have been set based on dose (e.g., permissible levels of lead in blood or mercury in urine). While these measures are more directly correlated with toxicity, it is still necessary to back-calculate exposure levels associated with these levels for purposes of controlling risks.

The next article concerns the factors and events that determine the relationships between exposure, dose and response. The first factors relate to uptake, absorption and distribution — the processes that determine the actual transport of substances into the body from the external environment across portals of entry such as skin, lung and gut. These processes are at the interface between humans and their environments. The second factors, of metabolism, relate to understanding how the body handles absorbed substances. Some substances are transformed by cellular processes of metabolism, which can either increase or decrease their biological activity.

The concepts of target organ and critical effect have been developed to aid in the interpretation of toxicological data. Depending upon dose, duration and route of exposure, as well as host factors such as age, many toxic agents can induce a number of effects within organs and organisms. An important role of toxicology is to identify the important effect or sets of effects in order to prevent irreversible or debilitating disease. One important part of this task is the identification of the organ first or most affected by a toxic agent; this organ is defined as the "target organ". Within the target organ, it is important to identify the important event or events that signals intoxication, or damage, in order to ascertain that the organ has been affected beyond the range of normal variation. This is known as the "critical effect"; it may represent the first event in a progression of pathophysiological stages (such as the excretion of small-molecular-weight proteins as a critical effect in nephrotoxicity), or it may represent the first and potentially irreversible effect in a disease process (such as formation of a DNA adduct in carcinogenesis). These concepts are important in occupational health because they define the types of toxicity and clinical disease associated with specific exposures, and in most cases reduction of exposure has as a goal the prevention of critical effects in target organs, rather than every effect in every or any organ.

The next two articles concern important host factors that affect many types of responses to many types of toxic agents. These are: genetic determinants, or inherited susceptibility/resistance factors; and age, sex and other factors such as diet or co-existence of infectious disease. These factors can also affect exposure and dose, through modifying uptake, absorption, distribution and metabolism. Because working populations around the world vary with respect to many of these factors, it is critical for occupational health specialists and policy-makers to understand the way in which these factors may contribute to variabilities in response among populations and individuals within populations. In societies with heterogeneous populations, these considerations are particularly important. The variability of human populations must be considered in evaluating the risks of occupational exposures and in reaching rational conclusions from the study of nonhuman organisms in toxicological research or testing.

The section then provides two general overviews on toxicology at the mechanistic level. Mechanistically, modern toxicologists consider that all toxic effects manifest their first actions at the cellular level; thus, cellular responses represent the earliest indications of the body's encounters with a toxic agent. It is further assumed that these responses represent a spectrum of events, from injury through death. Cell injury refers to specific processes utilized by cells, the smallest unit of biological organization within organs, to respond to challenge. These responses involve changes in the function of processes within the cell, including the membrane and its ability to take up, release or exclude substances; the directed synthesis of proteins from amino acids; and the turnover of cell components. These responses may be common to all injured cells, or they may be specific to certain types of cells within certain organ systems. Cell death is the destruction of cells within an organ system, as a consequence of irreversible or uncompensated cell injury. Toxic agents may cause cell death acutely because of certain actions such as poisoning oxygen transfer, or cell death may be the consequence of chronic intoxication. Cell death can be followed by replacement in some but not all organ systems, but in some conditions cell proliferation induced by cell death may be considered a toxic response. Even in the absence of cell death, repeated cell injury may induce stress within organs that compromises their function and affects their progeny.

The chapter is then divided into more specific topics, which are grouped into the following categories: mechanism, test methods, regulation and risk assessment. The mechanism articles mostly focus on target systems rather than organs. This reflects the practice of modern toxicology and medicine, which studies organ systems rather than isolated organs. Thus, for example, the discussion of genetic toxicology is not focused upon the toxic effects of agents within a specific organ but rather on genetic material as a target for toxic action. Likewise, the article on immunotoxicology discusses the various organs and cells of the immune system as targets for toxic agents. The methods articles are designed to be highly operational; they describe current methods in use in many countries for hazard identification, that is, the development of information related to biological properties of agents.

The chapter continues with five articles on the application of toxicology in regulation and policy-making, from hazard identification to risk assessment. The current practice in several countries, as well as IARC, is presented. These articles should enable the reader to understand how information derived from toxicology tests is integrated with basic and mechanistic inferences to derive quantitative information used in setting exposure levels and other approaches to controlling hazards in the workplace and general environment.

A summary of available toxicology databases, to which the readers of this encyclopaedia can refer for detailed information on specific toxic agents and exposures, can be found in Volume III (see "Toxicology databases" in the chapter Safe handling of chemicals, which provides information on many of these databases, their information sources, methods of evaluation and interpretation, and means of access). These databases, together with the Encyclopaedia, provide the occupational health specialist, the worker and the employer with the ability to obtain and use up-to-date information on toxicology and the evaluation of toxic agents by national and international bodies.

This chapter focuses upon those aspects of toxicology relevant to occupational safety and health. For that reason, clinical toxicology and forensic toxicology are not specifically addressed as subdisciplines of the field. Many of the same principles and approaches described here are used in these subdisciplines as well as in environmental health. They are also applicable to evaluating the impacts of toxic agents on nonhuman populations, a major concern of environmental policies in many countries. A committed attempt has been made to enlist the perspectives and experiences of experts and practitioners from all sectors and from many countries; however, the reader may note a certain bias towards academic scientists in the developed world. Although the editor and contributors believe that the principles and practice of toxicology are international, the problems of cultural bias and narrowness of experience may well be evident in this chapter. The chapter editor hopes that readers of this Encyclopaedia will assist in ensuring the broadest perspective possible as this important reference continues to be updated and expanded.

DEFINITIONS AND CONCEPTS

Exposure, Dose and Response

Toxicity is the intrinsic capacity of a chemical agent to affect an organism adversely.

Xenobiotics is a term for "foreign substances", that is, foreign to the organism. Its opposite is endogenous compounds. Xenobiotics include drugs, industrial chemicals, naturally occurring poisons and environmental pollutants.

Hazard is the potential for the toxicity to be realized in a specific setting or situation.

Risk is the probability of a specific adverse effect to occur. It is often expressed as the percentage of cases in a given population and during a specific time period. A risk estimate can be based upon actual cases or a projection of future cases, based upon extrapolations.

Toxicity rating and toxicity classification can be used for regulatory purposes. Toxicity rating is an arbitrary grading of doses or exposure levels causing toxic effects. The grading can be "supertoxic," "highly toxic," "moderately toxic" and so on. The most common ratings concern acute toxicity. Toxicity classification concerns the grouping of chemicals into general categories according to their most important toxic effect. Such categories can include allergenic, neurotoxic, carcinogenic and so on. This classification can be of administrative value as a warning and as information.

The dose-effect relationship is the relationship between dose and effect on the individual level. An increase in dose may increase the intensity of an effect, or a more severe effect may result. A dose-effect curve may be obtained at the level of the whole organism, the cell or the target molecule. Some toxic effects, such as death or cancer, are not graded but are "all or none" effects.

The dose-response relationship is the relationship between dose and the percentage of individuals showing a specific effect. With increasing dose a greater number of individuals in the exposed population will usually be affected.

It is essential to toxicology to establish dose-effect and dose-response relationships. In medical (epidemiological) studies a criterion often used for accepting a causal relationship between an agent and a disease is that effect or response is proportional to dose.

Several dose-response curves can be drawn for a chemical—one for each type of effect. The dose-response curve for most toxic effects (when studied in large populations) has a sigmoid shape. There is usually a low-dose range where there is no response detected; as dose increases, the response follows an ascending curve that will usually reach a plateau at a 100% response. The dose-response curve reflects the variations among individuals in a population. The slope of the curve varies from chemical to chemical and between different types of effects. For some chemicals with specific effects (carcinogens, initiators, mutagens) the dose-response curve might be linear from dose zero within a certain dose range. This means that no threshold exists and that even small doses represent a risk. Above that dose range, the risk may increase at greater than a linear rate.

Variation in exposure during the day and the total length of exposure during one's lifetime may be as important for the outcome (response) as mean or average or even integrated dose level. High peak exposures may be more harmful than a more even exposure level. This is the case for some organic solvents. On the other hand, for some carcinogens, it has been experimentally shown that the fractionation of a single dose into several exposures with the same total dose may be more effective in producing tumours.

A dose is often expressed as the amount of a xenobiotic entering an organism (in units such as mg/kg body weight). The dose may be expressed in different (more or less informative) ways: exposure dose, which is the air concentration of pollutant inhaled during a certain time period (in work hygiene usually eight hours), or the retained or absorbed dose (in industrial hygiene also called the body burden), which is the amount present in the body at a certain time during or after exposure. The tissue dose is the amount of substance in a specific tissue and the target dose is the amount of substance (usually a metabolite) bound to the critical molecule. The target dose can be expressed as mg chemical bound per mg of a specific macromolecule in the tissue. To apply this concept, information on the mechanism of toxic action on the molecular level is needed. The target dose is more exactly associated with the toxic effect. The exposure dose or body burden may be more easily available, but these are less precisely related to the effect.

In the dose concept a time aspect is often included, even if it is not always expressed. The theoretical dose according to Haber's law is $D = ct$, where D is dose, c is concentration of the xenobiotic in the air and t the duration of exposure to the chemical. If this concept is used at the target organ or molecular level, the amount per mg tissue or molecule over a certain time may be used. The time aspect is usually more important for understanding repeated exposures and chronic effects than for single exposures and acute effects.

Additive effects occur as a result of exposure to a combination of chemicals, where the individual toxicities are simply added to each other (1+1=2). When chemicals act *via* the same mechanism, additivity of their effects is assumed although not always the case in reality. Interaction between chemicals may result in an inhibition (antagonism), with a smaller effect than that expected from addition of the effects of the individual chemicals (1+1<2). Alternatively, a combination of chemicals may produce a more pronounced effect than would be expected by addition (increased response among individuals or an increase in frequency of response in a population), this is called synergism (1+1>2).

Latency time is the time between first exposure and the appearance of a detectable effect or response. The term is often used for carcinogenic effects, where tumours may appear a long time after the start of exposure and sometimes long after the cessation of exposure.

A dose threshold is a dose level below which no observable effect occurs. Thresholds are thought to exist for certain effects, like acute toxic effects; but not for others, like carcinogenic effects (by DNA-adduct-forming initiators). The mere absence of a response in a given population should not, however, be taken as evidence for the existence of a threshold. Absence of response could be due to simple statistical phenomena: an adverse effect occurring at low frequency may not be detectable in a small population.

LD_{50} (effective dose) is the dose causing 50% lethality in an animal population. The LD_{50} is often given in older literature as a measure of acute toxicity of chemicals. The higher the LD_{50}, the lower is the acute toxicity. A highly toxic chemical (with a low LD_{50}) is said to be potent. There is no necessary correlation between acute and chronic toxicity. ED_{50} (effective dose) is the dose causing a specific effect other than lethality in 50% of the animals.

NOEL (NOAEL) means the no observed (adverse) effect level, or the highest dose that does not cause a toxic effect. To establish a NOEL requires multiple doses, a large population and additional information to make sure that absence of a response is not merely a statistical phenomenon. LOEL is the lowest observed effective dose on a dose-response curve, or the lowest dose that causes an effect.

A safety factor is a formal, arbitrary number with which one divides the NOEL or LOEL derived from animal experiments to obtain a tentative permissible dose for humans. This is often used in the area of food toxicology, but may be used also in occupational toxicology. A safety factor may also be used for extrapolation of data from small populations to larger populations. Safety factors range from 10^0 to 10^3. A safety factor of two may typically be sufficient to protect from a less serious effect (such as irritation) and a factor as large as 1,000 may be used for very serious effects (such as cancer). The term safety factor could be better replaced by the term protection factor or, even, uncertainty factor. The use of the latter term reflects scientific uncertainties, such as whether exact dose-response data can be translated from animals to humans for the particular chemical, toxic effect or exposure situation.

Extrapolations are theoretical qualitative or quantitative estimates of toxicity (risk extrapolations) derived from translation of data from one species to another or from one set of dose-response data (typically in the high dose range) to regions of dose-response where no data exist. Extrapolations usually must be made to predict toxic responses outside the observation range. Mathematical modelling is used for extrapolations based upon an understanding of the behaviour of the chemical in the organism (toxicokinetic modelling) or based upon the understanding of statistical probabilities that specific biological events will occur (biologically or mechanistically based models). Some national agencies have developed sophisticated extrapolation models as a formalized method to predict risks for regulatory purposes. (See discussion of risk assessment later in the chapter.)

Systemic effects are toxic effects in tissues distant from the route of absorption.

Target organ is the primary or most sensitive organ affected after exposure. The same chemical entering the body by different routes of exposure dose, dose rate, sex and species may affect different target organs. Interaction between chemicals, or between chemicals and other factors may affect different target organs as well.

Acute effects occur after limited exposure and shortly (hours, days) after exposure and may be reversible or irreversible.

Chronic effects occur after prolonged exposure (months, years, decades) and/ or persist after exposure has ceased.

Acute exposure is an exposure of short duration, while chronic exposure is long-term (sometimes life-long) exposure.

Tolerance to a chemical may occur when repeat exposures result in a lower response than what would have been expected without pretreatment.

UPTAKE AND DISPOSITION

Transport Processes

Diffusion. In order to enter the organism and reach a site where damage is produced, a foreign substance has to pass several barriers, including cells and their membranes. Most toxic substances pass through membranes passively by diffusion. This may occur for small water-soluble molecules by passage through aqueous channels or, for fat-soluble ones, by dissolution into and diffusion through the lipid part of the membrane. Ethanol, a small molecule that is both water and fat soluble, diffuses rapidly through cell membranes.

Diffusion of weak acids and bases. Weak acids and bases may readily pass membranes in their non-ionized, fat-soluble form while ionized forms are too polar to pass. The degree of ionization of these substances depends on pH. If a pH gradient exists across a membrane they will therefore accumulate on one side. The urinary excretion of weak acids and bases is highly dependent on urinary pH. Foetal or embryonic pH is somewhat higher than maternal pH, causing a slight accumulation of weak acids in the foetus or embryo.

Facilitated diffusion. The passage of a substance may be facilitated by carriers in the membrane. Facilitated diffusion is similar to enzyme processes in that it is protein mediated, highly selective, and saturable. Other substances may inhibit the facilitated transport of xenobiotics.

Active transport. Some substances are actively transported across cell membranes. This transport is mediated by carrier proteins in a process analogous to that of enzymes. Active transport is similar to facilitated diffusion, but it may occur against a concentration gradient. It requires energy input and a metabolic inhibitor can block the process. Most environmental pollutants are not transported actively. One exception is the active tubular secretion and reabsorption of acid metabolites in the kidneys.

Phagocytosis is a process where specialized cells such as macrophages engulf particles for subsequent digestion. This transport process is important, for example, for the removal of particles in the alveoli.

Bulk flow. Substances are also transported in the body along with the movement of air in the respiratory system during breathing, and the movements of blood, lymph or urine.

Filtration. Due to hydrostatic or osmotic pressure water flows in bulk through pores in the endothelium. Any solute that is small enough will be filtered together with the water. Filtration occurs to some extent in the capillary bed in all tissues but is particularly important in the formation of primary urine in the kidney glomeruli.

Absorption

Absorption is the uptake of a substance from the environment into the organism. The term usually includes not only the entrance into the barrier tissue but also the further transport into circulating blood.

Pulmonary absorption. The lungs are the primary route of deposition and absorption of small airborne particles, gases, vapours and aerosols. For highly water-soluble gases and vapours a significant part of the uptake occurs in the nose and the respiratory tree, but for less soluble substances it primarily takes place in the lung alveoli. The alveoli have a very large surface area (about 100 m^2 in humans). In addition, the diffusion barrier is extremely small, with only two thin cell layers and a distance in the order of micrometers from alveolar air to systemic blood circulation.

This makes the lungs very efficient not only in the exchange of oxygen and carbon dioxide but also of other gases and vapours. In general, the diffusion across the alveolar wall is so rapid that it does not limit the uptake. The absorption rate is instead dependent on flow (pulmonary ventilation, cardiac output) and solubility (blood: air partition coefficient). Another important factor is metabolic elimination. The relative importance of these factors for pulmonary absorption varies greatly for different substances. Physical activity results in increased pulmonary ventilation and cardiac output, and decreased liver blood flow (and, hence, biotransformation rate). For many inhaled substances this leads to a marked increase in pulmonary absorption.

Percutaneous absorption. The skin is a very efficient barrier. Apart from its thermoregulatory role, it is designed to protect the organism from microorganisms, ultraviolet radiation and other deleterious agents, and also against excessive water loss. The diffusion distance in the dermis is on the order of tenths of millimetres. In addition, the keratin layer has a very high resistance to diffusion for most substances. Nevertheless, significant dermal absorption resulting in toxicity may occur for some substances – highly toxic, fat-soluble substances such as organophosphorous insecticides and organic solvents, for example. Significant absorption is likely to occur after exposure to liquid substances. Percutaneous absorption of vapour may be important for solvents with very low vapour pressure and high affinity to water and skin.

Gastrointestinal absorption occurs after accidental or intentional ingestion. Larger particles originally inhaled and deposited in the respiratory tract may be swallowed after mucociliary transport to the pharynx. Practically all soluble substances are efficiently absorbed in the gastrointestinal tract. The low pH of the gut may facilitate absorption, for instance, of metals.

Other routes. In toxicity testing and other experiments, special routes of administration are often used for convenience, although these are rare and usually not relevant in the occupational setting. These routes include intravenous (IV), subcutaneous (sc), intraperitoneal (ip) and intramuscular (im) injections. In general, substances are absorbed at a higher rate and more completely by these routes, especially after IV injection. This leads to short-lasting but high concentration peaks that may increase the toxicity of a dose.

Distribution

The distribution of a substance within the organism is a dynamic process which depends on uptake and elimination rates, as well as the blood flow to the different tissues and their affinities for the substance. Water-soluble, small, uncharged molecules, univalent cations, and most anions diffuse easily and will eventually reach a relatively even distribution in the body.

Volume of distribution is the amount of a substance in the body at a given time, divided by the concentration in blood, plasma or serum at that time. The value has no meaning as a physical volume, as many substances are not uniformly distributed in the organism. A volume of distribution of less than one l/kg body weight indicates preferential distribution in the blood (or serum or plasma), whereas a value above one indicates a preference for peripheral tissues such as adipose tissue for fat soluble substances.

Accumulation is the build-up of a substance in a tissue or organ to higher levels than in blood or plasma. It may also refer to a gradual build-up over time in the organism. Many xenobiotics are highly fat soluble and tend to accumulate in adipose tissue, while others have a special affinity for bone. For example, calcium in bone may be exchanged for cations of lead, strontium, barium and radium, and hydroxyl groups in bone may be exchanged for fluoride.

Barriers. The blood vessels in the brain, testes and placenta have special anatomical features that inhibit passage of large molecules like proteins. These features, often referred to as blood-brain, blood-testes, and blood-placenta barriers, may give the false impression that they prevent passage of any substance. These barriers are of little or no importance for xenobiotics that can diffuse through cell membranes.

Blood binding. Substances may be bound to red blood cells or plasma components, or occur unbound in blood. Carbon monoxide, arsenic, organic mercury and hexavalent chromium have a high affinity for red blood cells, while inorganic mercury and trivalent chromium show a preference for plasma proteins. A number of other substances also bind to plasma proteins. Only the unbound fraction is available for filtration or diffusion into eliminating organs. Blood binding may therefore increase the residence time in the organism but decrease uptake by target organs.

Elimination

Elimination is the disappearance of a substance in the body. Elimination may involve excretion from the body or transformation to other substances not captured by a specific method of measurement. The rate of disappearance may be expressed by the elimination rate constant, biological half-time or clearance.

Concentration-time curve. The curve of concentration in blood (or plasma) versus time is a convenient way of describing uptake and disposition of a xeno-biotic.

Area under the curve (AUC) is the integral of concentration in blood (plasma) over time. When metabolic saturation and other non-linear processes are absent, AUC is proportional to the absorbed amount of substance.

Biological half-time (or half-life) is the time needed after the end of exposure to reduce the amount in the organism to one-half. As it is often difficult to assess the total amount of a substance, measurements such as the concentration in blood (plasma) are used. The half-time should be used with caution, as it may change, for example, with dose and length of exposure. In addition, many substances have complex decay curves with several half-times.

Bioavailability is the fraction of an administered dose entering the systemic circulation. In the absence of presystemic clearance, or first-pass metabolism, the fraction is one. In oral exposure presystemic clearance may be due to metabolism within the gastrointestinal content, gut wall or liver. First-pass metabolism will reduce the systemic absorption of the substance and instead increase the absorption of metabolites. This may lead to a different toxicity pattern.

Clearance is the volume of blood (plasma) per unit time completely cleared of a substance. To distinguish from renal clearance, for example, the prefix total, metabolic or blood (plasma) is often added.

Intrinsic clearance is the capacity of endogenous enzymes to transform a substance, and is also expressed in volume per unit time. If the intrinsic clearance in an organ is much lower than the blood flow, the metabolism is said to be capacity limited. Conversely, if the intrinsic clearance is much higher than the blood flow, the metabolism is flow limited.

Excretion

Excretion is the exit of a substance and its biotransformation products from the organism.

Excretion in urine and bile. The kidneys are the most important excretory organs. Some substances, especially acids with high molecular weights, are excreted with bile. A fraction of biliary excreted substances may be reabsorbed in the intestines. This process, enterohepatic circulation, is common for conjugated substances following intestinal hydrolysis of the conjugate.

Other routes of excretion. Some substances, such as organic solvents and breakdown products such as acetone, are volatile enough so that a considerable fraction may be excreted by exhalation after inhalation. Small water-soluble molecules as well as fat-soluble ones are readily secreted to the foetus *via* the placenta, and into milk in mammals. For the mother, lactation can be a quantitatively important excretory pathway for persistent fat-soluble chemicals. The offspring may be secondarily exposed *via* the mother during pregnancy as well as during lactation. Water-soluble compounds may to some extent be excreted in sweat and saliva. These routes are generally of minor importance. However, as a large volume of saliva is produced and swallowed, saliva excretion may contribute to reabsorption of the compound. Some metals such as mercury are excreted by binding permanently to the sulphydryl groups of the keratin in the hair.

Toxicokinetic Models

Mathematical models are important tools to understand and describe the uptake and disposition of foreign substances. Most models are compartmental, that is, the organism is represented by one or more compartments. A compartment is a chemically and physically theoretical volume in which the substance is assumed to distribute homogeneously and instantaneously. Simple models may be expressed as a sum of exponential terms, while more complicated ones require numerical procedures on a computer for their solution. Models may be subdivided in two categories, descriptive and physiological.

In descriptive models, fitting to measured data is performed by changing the numerical values of the model parameters or even the model structure itself. The model structure normally has little to do with the structure of the organism. Advantages of the descriptive approach are that few assumptions are made and that there is no need for additional data. A disadvantage of descriptive models is their limited usefulness for extrapolations.

Physiological models are constructed from physiological, anatomical and other independent data. The model is then refined and validated by comparison with experimental data. An advantage of physiological models is that they can be used for extrapolation purposes. For example, the influence of physical activity on the uptake and disposition of inhaled substances may be predicted from known physiological adjustments in ventilation and cardiac output. A disadvantage of physiological models is that they require a large amount of independent data.

Biotransformation

Biotransformation is a process which leads to a metabolic conversion of foreign compounds (xenobiotics) in the body. The process is often referred to as metabolism of xenobiotics. As a general rule metabolism converts lipid-soluble xenobiotics to large, watersoluble metabolites that can be effectively excreted.

The liver is the main site of biotransformation. All xenobiotics taken up from the intestine are transported to the liver by a single blood vessel (vena porta). If taken up in small quantities a foreign substance may be completely metabolized in the liver before reaching the general circulation and other organs (first pass effect). Inhaled xenobiotics are distributed *via* the general circulation to the liver. In that case only a fraction of the dose is metabolized in the liver before reaching other organs.

Liver cells contain several enzymes that oxidize xenobiotics. This oxidation generally activates the compound — it becomes more reactive than the parent molecule. In most cases the oxidized metabolite is further metabolized by other enzymes in a second phase. These enzymes conjugate the metabolite with an endogenous substrate, so that the molecule becomes larger and more polar. This facilitates excretion.

Enzymes that metabolize xenobiotics are also present in other organs such as the lungs and kidneys. In these organs they may play specific and qualitatively important roles in the metabolism of certain xenobiotics. Metabolites formed in one organ may be further metabolized in a second organ. Bacteria in the intestine may also participate in biotransformation.

Metabolites of xenobiotics can be excreted by the kidneys or *via* the bile. They can also be exhaled *via* the lungs, or bound to endogenous molecules in the body.

The relationship between biotransformation and toxicity is complex. Biotransformation can be seen as a necessary process for survival. It protects the organism against toxicity by preventing accumulation of harmful substances in the body. However, reactive intermediary metabolites may be formed in biotransformation, and these are potentially harmful. This is called metabolic activation. Thus, biotransformation may also induce toxicity. Oxidized, intermediary metabolites that are not conjugated can bind to and damage cellular structures. If, for example, a xenobiotic metabolite binds to DNA, a mutation can be induced (see "Genetic toxicology"). If the biotransformation system is overloaded, a massive destruction of essential proteins or lipid membranes may occur. This can result in cell death (see "Cellular injury and cellular death").

Metabolism is a word often used interchangeably with biotransformation. It denotes chemical breakdown or synthesis reactions catalyzed by enzymes in the body. Nutrients from food, endogenous compounds, and xenobiotics are all metabolized in the body.

Metabolic activation means that a less reactive compound is converted to a more reactive molecule. This usually occurs during Phase 1 reactions.

Metabolic inactivation means that an active or toxic molecule is converted to a less active metabolite. This usually occurs during Phase 2 reactions. In certain cases an inactivated metabolite might be reactivated, for example by enzymatic cleavage.

Phase 1 reaction refers to the first step in xenobiotic metabolism. It usually means that the compound is oxidized. Oxidation usually makes the compound more water soluble and facilitates further reactions.

Cytochrome P450 enzymes are a group of enzymes that preferentially oxidize xenobiotics in Phase 1 reactions. The different enzymes are specialized for handling specific groups of xenobiotics with certain characteristics. Endogenous molecules are also substrates. Cytochrome P450 enzymes are induced by xenobiotics in a specific fashion. Obtaining induction data on cytochrome P450 can be informative about the nature of previous exposures (see "Genetic determinants of toxic response").

Phase 2 reaction refers to the second step in xenobiotic metabolism. It usually means that the oxidized compound is conjugated with (coupled to) an endogenous molecule. This reaction increases the water solubility further. Many conjugated metabolites are actively excreted *via* the kidneys.

Transferases are a group of enzymes that catalyze Phase 2 reactions. They conjugate xenobiotics with endogenous compounds such as glutathione, amino acids, glucuronic acid or sulphate.

Glutathione is an endogenous molecule, a tripeptide, that is conjugated with xenobiotics in Phase 2 reactions. It is present in all cells (and in liver cells in high concentrations), and usually protects from activated xenobiotics. When glutathione is depleted, toxic reactions between activated xenobiotic metabolites and proteins, lipids or DNA may occur.

Induction means that enzymes involved in biotransformation are increased (in activity or amount) as a response to xenobiotic exposure. In some cases within a few days enzyme activity can be increased several fold. Induction is often balanced so that both Phase 1 and Phase 2 reactions are increased simultaneously. This may lead to a more rapid biotransformation and can explain tolerance. In contrast, unbalanced induction may increase toxicity.

Inhibition of biotransformation can occur if two xenobiotics are metabolized by the same enzyme. The two substrates have to compete, and usually one of the substrates is preferred. In that case the second substrate is not metabolized, or only slowly metabolized. As with induction, inhibition may increase as well as decrease toxicity.

Oxygen activation can be triggered by metabolites of certain xenobiotics. They may auto-oxidize under the production of activated oxygen species. These oxygen-derived species, which include superoxide, hydrogen peroxide and the hydroxyl radical, may damage DNA, lipids and proteins in cells. Oxygen activation is also involved in inflammatory processes.

Genetic variability between individuals is seen in many genes coding for Phase 1 and Phase 2 enzymes. Genetic variability may explain why certain individuals are more susceptible to toxic effects of xenobiotics than others.

TOXICOKINETICS

The human organism represents a complex biological system on various levels of organization, from the molecular-cellular level to the tissues and organs. The organism is an open system, exchanging matter and energy with the environment through numerous biochemical reactions in a dynamic equilibrium. The environment can be polluted, or contaminated with various toxicants.

Penetration of molecules or ions of toxicants from the work or living environment into such a strongly coordinated biological system can reversibly or irreversibly disturb normal cellular biochemical processes, or even injure and destroy the cell (see "Cellular injury and cellular death").

Penetration of a toxicant from the environment to the sites of its toxic effect inside the organism can be divided into three phases:

1. The exposure phase encompasses all processes occurring between various toxicants and/or the influence on them of environmental factors (light, temperature, humidity, *etc.*). Chemical transformations, degradation, biodegradation (by micro-organisms) as well as disintegration of toxicants can occur.

2. The toxicokinetic phase encompasses absorption of toxicants into the organism and all processes which follow: transport by body fluids, distribution and accumulation in tissues and organs, biotransformation to metabolites and elimination (excretion) of toxicants and/or metabolites from the organism.

3. The toxicodynamic phase refers to the interaction of toxicants (molecules, ions, colloids) with specific sites of action on or inside the cells—receptors—ultimately producing a toxic effect.

Here we will focus our attention exclusively on the toxicokinetic processes inside the human organism following exposure to toxicants in the environment.

The molecules or ions of toxicants present in the environment will penetrate into the organism through the skin and mucosa, or the epithelial cells of the respiratory and gastrointestinal tracts, depending on the point of entry. That means molecules and ions of toxicants must penetrate through cellular membranes of these biological systems, as well as through an intricate system of endomembranes inside the cell.

All toxicokinetic and toxicodynamic processes occur on the molecular-cellular level. Numerous factors influence these processes and these can be divided into two basic groups:

*   chemical constitution and physicochemical properties of toxicants
*   structure of the cell especially properties and function of membranes around the cell and its interior organelles.

Physico-Chemical Properties of Toxicants

In 1854 the Russian toxicologist E.V. Pelikan started studies on the relation between the chemical structure of a substance and its biological activity—the structure activity relationship (SAR). Chemical structure directly determines physico-chemical properties, some of which are responsible for biological activity.

To define the chemical structure numerous parameters can be selected as descriptors, which can be divided into various groups:

1. Physico-chemical:
   - general — melting point, boiling point, vapour pressure, dissociation constant (pK_a), Nernst partition coefficient (P), activation energy, heat of reaction, reduction potential, *etc.*
   - electric — ionization potential, dielectric constant, dipole moment, mass: charge ratio, *etc.*
   - quantum chemical — atomic charge, bond energy, resonance energy, electron density, molecular reactivity, *etc.*
1. Steric: molecular volume, shape and surface area, substructure shape, molecular reactivity, *etc.*
2. Structura: number of bonds number of rings (in polycyclic compounds), extent of branching, *etc.*

For each toxicant it is necessary to select a set of descriptors related to a particular mechanism of activity. However, from the toxicokinetic point of view two parameters are of general importance for all toxicants:

- The Nernst partition coefficient (P) establishes the solubility of toxicant molecules in the two-phase octanol (oil)-water system, correlating to their lipo- or hydrosolubility. This parameter will greatly influence the distribution and accumulation of toxicant molecules in the organism.

- The dissociation constant (pK_a) defines the degree of ionization (electrolytic dissociation) of molecules of a toxicant into charged cations and anions at a particular pH. This constant represents the pH at which 50% ionization is achieved. Molecules can be lipophilic or hydrophilic, but ions are soluble exclusively in the water of body fluids and tissues. Knowing pK_a it is possible to calculate the degree of ionization of a substance for each pH using the Henderson-Hasselbach equation.

For inhaled dusts and aerosols, the particle size, shape, surface area and density also influence their toxicokinetics and toxicodynamics.

Structure and Properties of Membranes

The eukaryotic cell of human and animal organisms is encircled by a cytoplasmic membrane regulating the transport of substances and maintaining cell homeostasis. The cell organelles (nucleus, mitochondria) possess membranes too. The cell cytoplasm is compartmentalized by intricate membranous structures, the endoplasmic reticulum and Golgi complex (endomembranes). All these membranes are structurally alike, but vary in the content of lipids and proteins.

The structural framework of membranes is a bilayer of lipid molecules (phospholipids, sphyngolipids, cholesterol). The backbone of a phospholipid molecule is glycerol with two of its -OH groups esterified by aliphatic fatty acids with 16 to 18 carbon atoms, and the third group esterified by a phosphate group and a nitrogenous compound (choline, ethanolamine, serine). In sphyngolipids, sphyngosine is the base.

The lipid molecule is amphipatic because it consists of a polar hydrophilic "head" (amino alcohol, phosphate, glycerol) and a non-polar twin "tail" (fatty acids). The lipid bilayer is arranged so that the hydrophilic heads constitute the outer and inner surface of membrane and lipophilic tails are stretched toward the membrane interior, which contains water, various ions and molecules.

Proteins and glycoproteins are inserted into the lipid bilayer (intrinsic proteins) or attached to the membrane surface (extrinsic proteins). These proteins contribute to the structural integrity of the membrane, but they may also perform as enzymes, carriers, pore walls or receptors.

The membrane represents a dynamic structure which can be disintegrated and rebuilt with a different proportion of lipids and proteins, according to functional needs.

Regulation of transport of substances into and out of the cell represents one of the basic functions of outer and inner membranes.

Some lipophilic molecules pass directly through the lipid bilayer. Hydrophilic molecules and ions are transported *via* pores. Membranes respond to changing conditions by opening or sealing certain pores of various sizes.

The following processes and mechanisms are involved in the transport of substances, including toxicants, through membranes:

- diffusion through lipid bilayer
- diffusion through pores
- transport by a carrier (facilitated diffusion).

Active processes:

- active transport by a carrier
- endocytosis (pinocytosis).

Diffusion

This represents the movement of molecules and ions through lipid bilayer or pores from a region of high concentration, or high electric potential, to a region of low concentration or potential ("downhill"). Difference in concentration or electric charge is the driving force influencing the intensity of the flux in both directions. In the equilibrium state, influx will be equal to efflux. The rate of diffusion follows Ficke's law, stating that it is directly proportional to the available surface of membrane, difference in concentration (charge) gradient and characteristic diffusion coefficient, and inversely proportional to the membrane thickness.

Small lipophilic molecules pass easily through the lipid layer of membrane, according to the Nernst partition coefficient.

Large lipophilic molecules, water soluble molecules and ions will use aqueous pore channels for their passage. Size and stereoconfiguration will influence passage of molecules. For ions, besides size, the type of charge will be decisive. The protein molecules of pore walls can gain positive or negative charge. Narrow pores tend to be selective — negatively charged ligands will allow passage only for cations, and positively charged ligands will allow passage only for anions. With the increase of pore diameter hydrodynamic flow is dominant, allowing free passage of ions and molecules, according to Poiseuille's law. This filtration is a consequence of the osmotic gradient. In some cases ions can penetrate through specific complex molecules — ionophores — which can be produced by micro-organisms with anti-biotic effects (nonactin, valinomycin, gramacidin, *etc.*).

Facilitated or Catalyzed Diffusion

This requires the presence of a carrier in the membrane, usually a protein molecule (permease). The carrier selectively binds substances, resembling a substrate-enzyme complex. Similar molecules (including toxicants) can compete for the specific carrier until its saturation point is reached. Toxicants can compete for the carrier and when they are irreversibly bound to it the transport is blocked. The rate of transport is characteristic for each type of carrier. If transport is performed in both direction, it is called exchange diffusion.

Active Transport

For transport of some substances vital for the cell, a special type of carrier is used, transporting against the concentration gradient or electric potential ("up-hill"). The carrier is very stereospecific and can be saturated.

For uphill transport, energy is required. The necessary energy is obtained by catalytic cleavage of ATP molecules to ADP by the enzyme adenosine triphos-phatase (ATP-ase).

Toxicants can interfere with this transport by competitive or non-competitive inhibition of the carrier or by inhibition of ATP-ase activity.

Endocytosis

Endocytosis is defined as a transport mechanism in which the cell membrane encircles material by enfolding to form a vesicle transporting it through the cell. When the material is liquid, the process is termed pinocytosis. In some cases the material is bound to a receptor and this complex is transported by a membrane vesicle. This type of transport is especially used by epithelial cells of the gastro-intestinal tract, and cells of the liver and kidneys.

Absorption of Toxicants

People are exposed to numerous toxicants present in the work and living environment, which can penetrate into the human organism by three main portals of entry:

- *via* the respiratory tract by inhalation of polluted air

- *via* the gastrointestinal tract by ingestion of contaminated food, water and drinks
- through the skin by dermal, cutaneous penetration.

In the case of exposure in industry, inhalation represents the dominant way of entry of toxicants, followed by dermal penetration. In agriculture, pesticides exposure *via* dermal absorption is almost equal to cases of combined inhalation and dermal penetration. The general population is mostly exposed by ingestion of contaminated food, water and beverages, then by inhalation and less often by dermal penetration.

Absorption *via* the Respiratory Tract

Absorption in the lungs represents the main route of uptake for numerous airborne toxicants (gases, vapours, fumes, mists, smokes, dusts, aerosols, *etc.*).

The respiratory tract (RT) represents an ideal gas-exchange system possessing a membrane with a surface of 30 m^2 (expiration) to 100 m^2 (deep inspiration), behind which a network of about 2,000 km of capillaries is located. The system, developed through evolution, is accommodated into a relatively small space (chest cavity) protected by ribs.

Anatomically and physiologically the RT can be divided into three compartments:

- the upper part of RT, or nasopharyngeal (NP), starting at nose nares and extended to the pharynx and larynx; this part serves as an air-conditioning system
- the tracheo-bronchial tree (TB), encompassing numerous tubes of various sizes, which bring air to the lungs
- the pulmonary compartment (P), which consists of millions of alveoli (air-sacs) arranged in grapelike clusters.

Hydrophilic toxicants are easily absorbed by the epithelium of the nasopharyngeal region. The whole epithelium of the NP and TB regions is covered by a film of water. Lipophilic toxicants are partially absorbed in the NP and TB, but mostly in the alveoli by diffusion through alveolo-capillary membranes. The absorption rate depends on lung ventilation, cardiac output (blood flow through lungs), solubility of toxicant in blood and its metabolic rate.

In the alveoli, gas exchange is carried out. The alveolar wall is made up of an epithelium, an interstitial framework of basement membrane, connective tissue and the capillary endothelium. The diffusion of toxicants is very rapid through these layers, which have a thickness of about 0.8 μm. In alveoli, toxicant is transferred from the air phase into the liquid phase (blood). The rate of absorption (air to blood distribution) of a toxicant depends on its concentration in alveolar air and the Nernst partition coefficient for blood (solubility coefficient).

In the blood the toxicant can be dissolved in the liquid phase by simple physical processes or bound to the blood cells and/or plasma constituents according to chemical affinity or by adsorption. The water content of blood is 75% and, therefore, hydrophilic gases and vapours show a high solubility in plasma (*e.g.*, alcohols). Lipophilic toxicants (*e.g.*, benzene) are usually bound to cells or macro-molecules such as albumen.

From the very beginning of exposure in the lungs, two opposite processes are occurring: absorption and desorption. The equilibrium between these processes depends on the concentration of toxicant in alveolar air and blood. At the onset of exposure the toxicant concentration in the blood is 0 and retention in blood is almost 100%. With continuation of exposure, an equilibrium between absorption and desorption is attained. Hydrophilic toxicants will rapidly attain equilibrium, and the rate of absorption depends on pulmonary ventilation rather than on blood flow. Lipophilic toxicants need a longer time to achieve equilibrium, and here the flow of unsaturated blood governs the rate of absorption.

Deposition of particles and aerosols in the RT depends on physical and physiological factors, as well as particle size. In short, the smaller the particle the deeper it will penetrate into the RT.

Relatively constant low retention of dust particles in the lungs of persons who are highly exposed (*e.g.*, miners) suggests the existence of a very efficient system for the clearance of particles. In the upper part of the RT (tracheo-bronchial) a mucociliary blanket performs the clearance. In the pulmonary part, three different mechanisms are at work. : (1) mucociliary blanket, (2) phagocytosis and (3) direct penetration of particles through the alveolar wall.

The first 17 of the 23 branchings of the tracheo-bronchial tree possess ciliated epithelial cells. By their strokes these cilia constantly move a mucous blanket toward the mouth. Particles deposited on this mucociliary blanket will be swallowed in the mouth (ingestion). A mucous blanket also covers the surface of the alveolar epithelium, moving toward the mucociliary blanket. Additionally the specialized moving cells — phagocytes — engulf particles and micro-organisms in the alveoli and migrate in two possible directions:

- toward the mucociliary blanket, which transports them to the mouth
- through the intercellular spaces of the alveolar wall to the lymphatic system of the lungs; also particles can directly penetrate by this route.

Absorption *via* Gastrointestinal Tract

Toxicants can be ingested in the case of accidental swallowing, intake of contaminated food and drinks, or swallowing of particles cleared from the RT.

The entire alimentary channel, from oesophagus to anus, is basically built in the same way. A mucous layer (epithelium) is supported by connective tissue and then by a network of capillaries and smooth muscle. The surface epithelium of the stomach is very wrinkled to increase the absorption/secretion surface area. The intestinal area contains numerous small projections (villi), which are able to absorb material by "pumping in". The active area for absorption in the intestines is about 100 m^2.

In the gastrointestinal tract (GIT) all absorption processes are very active:

- transcellular transport by diffusion through the lipid layer and/or pores of cell membranes, as well as pore filtration
- paracellular diffusion through junctions between cells
- facilitated diffusion and active transport
- endocytosis and the pumping mechanism of the villi.

Some toxic metal ions use specialized transport systems for essential elements: thallium, cobalt and manganese use the iron system, while lead appears to use the calcium system.

Many factors influence the rate of absorption of toxicants in various parts of the GIT:

- physico-chemical properties of toxicants, especially the Nernst partition coefficient and the dissociation constant; for particles, particle size is important—the smaller the size, the higher the solubility
- quantity of food present in the GIT (diluting effect)
- residence time in each part of the GIT (from a few minutes in the mouth to one hour in the stomach to many hours in the intestines
- the absorption area and absorption capacity of the epithelium
- local pH, which governs absorption of dissociated toxicants; in the acid pH of the stomach, non-dissociated acidic compounds will be more quickly absorbed
- peristalsis (movement of intestines by muscles) and local blood flow
- gastric and intestinal secretions transform toxicants into more or less soluble products; bile is an emulsifying agent producing more soluble complexes (hydrotrophy)
- combined exposure to other toxicants, which can produce synergistic or antagonistic effects in absorption processes
- presence of complexing/chelating agents
- the action of microflora of the RT (about 1.5 kg), about 60 different bacterial species which can perform biotransformation of toxicants.

It is also necessary to mention the enterohepatic circulation. Polar toxicants and/or metabolites (glucuronides and other conjugates) are excreted with the bile into the duodenum. Here the enzymes of the microflora perform hydrolysis and liberated products can be reabsorbed and transported by the portal vein into the liver. This mechanism is very dangerous in the case of hepatotoxic substances, enabling their temporary accumulation in the liver.

In the case of toxicants biotransformed in the liver to less toxic or non-toxic metabolites, ingestion may represent a less dangerous portal of entry. After absorption in the GIT these toxicants will be transported by the portal vein to the liver, and there they can be partially detoxified by biotransformation.

Absorption through the Skin (Dermal, Percutaneous)

The skin (1.8 m^2 of surface in a human adult) together with the mucous membranes of the body orifices, covers the surface of the body. It represents a barrier against physical, chemical and biological agents, maintaining the body integrity and homeostasis and performing many other physiological tasks.

Basically the skin consists of three layers: epidermis, true skin (dermis) and subcutaneous tissue (hypodermis). From the toxicological point of view the epidermis is of most interest here. It is built of many layers of cells. A horny surface of flattened, dead cells (stratum corneum) is the top layer, under which a continuous layer of living cells (stratum corneum compactum) is located, followed by a typical lipid membrane, and then by stratum lucidum, stratum gramulosum and stratum mucosum. The lipid membrane represents a protective barrier, but in hairy parts of the skin, both hair follicles and sweat gland channels penetrate through it. Therefore, dermal absorption can occur by the following mechanisms:

- transepidermal absorption by diffusion through the lipid membrane (barrier), mostly by lipophilic substances (organic solvents, pesticides, *etc.*) and to a small extent by some hydrophilic substances through pores
- transfollicular absorption around the hair stalk into the hair follicle, bypassing the membrane barrier; this absorption occurs only in hairy areas of skin
- absorption *via* the ducts of sweat glands, which have a cross-sectional area of about 0.1 to 1% of the total skin area (relative absorption is in this proportion)
- absorption through skin when injured mechanically, thermally, chemically or by skin diseases; here the skin layers, including lipid barrier, are disrupted and the way is open for toxicants and harmful agents to enter.

The rate of absorption through the skin will depend on many factors:

- concentration of toxicant, type of vehicle (medium), presence of other substances
- water content of skin, pH, temperature, local blood flow, perspiration, surface area of contaminated skin, thickness of skin
- anatomical and physiological characteristics of the skin due to sex, age, individual variations, differences occurring in various ethnic groups and races, *etc.*

Transport of Toxicants by Blood and Lymph

After absorption by any of these portals of entry, toxicants will reach the blood, lymph or other body fluids. The blood represents the major vehicle for transport of toxicants and their metabolites.

Blood is a fluid circulating organ, transporting necessary oxygen and vital substances to the cells and removing waste products of metabolism. Blood also contains cellular components, hormones, and other molecules involved in many physiological functions. Blood flows inside a relatively well closed, high-pressure circulatory system of blood vessels, pushed by the activity of the heart. Due to high pressure, leakage of fluid occurs. The lymphatic system represents the drainage system, in the form of a fine mesh of small, thin-walled lymph capillaries branching through the soft tissues and organs.

Blood is a mixture of a liquid phase (plasma, 55%) and solid blood cells (45%). Plasma contains proteins (albumins, globulins, fibrinogen), organic acids (lactic, glutamic, citric) and many other substances (lipids, lipoproteins, glycoproteins, enzymes, salts, xenobiotics, *etc.*). Blood cell elements include erythrocytes (Er), leukocytes, reticulocytes, monocytes, and platelets.

Toxicants are absorbed as molecules and ions. Some toxicants at blood pH form colloid particles as a third form in this liquid. Molecules, ions and colloids of toxicants have various possibilities for transport in blood:

- to be physically or chemically bound to the blood elements, mostly Er
- to be physically dissolved in plasma in a free state
- to be bound to one or more types of plasma proteins, complexed with the organic acids or attached to other fractions of plasma.

Most of the toxicants in blood exist partially in a free state in plasma and partially bound to erythrocytes and plasma constituents. The distribution depends on the affinity of toxicants to these constituents. All fractions are in a dynamic equilibrium.

Some toxicants are transported by the blood elements — mostly by erythrocytes, very rarely by leukocytes. Toxicants can be adsorbed on the surface of Er, or can bind to the ligands of stroma. If they penetrate into Er they can bind to the haem (*e.g.* carbon monoxide and selenium) or to the globin (Sb^{111}, Po^{210}). Some toxicants transported by Er are arsenic, cesium, thorium, radon, lead and sodium. Hexavalent chromium is exclusively bound to the Er and trivalent chromium to the proteins of plasma. For zinc, competition between Er and plasma occurs. About 96% of lead is transported by Er. Organic mercury is mostly bound to Er and inorganic mercury is carried mostly by plasma albumin. Small fractions of beryllium, copper, tellurium and uranium are carried by Er.

The majority of toxicants are transported by plasma or plasma proteins. Many electrolytes are present as ions in an equilibrium with non-dissociated molecules free or bound to the plasma fractions. This ionic fraction of toxicants is very diffusible, penetrating through the walls of capillaries into tissues and organs. Gases and vapours can be dissolved in the plasma.

Plasma proteins possess a total surface area of about 600 to 800 km^2 offered for absorption of toxicants. Albumin molecules possess about 109 cationic and 120 anionic ligands at the disposal of ions. Many ions are partially carried by albumin (*e.g.*, copper, zinc and cadmium), as are such compounds as dinitro- and ortho-cresols, nitro- and halogenated derivatives of aromatic hydrocarbons, and phenols.

Globulin molecules (alpha and beta) transport small molecules of toxicants as well as some metallic ions (copper, zinc and iron) and colloid particles. Fibrinogen shows affinity for certain small molecules. Many types of bonds can be involved in binding of toxicants to plasma proteins: Van der Waals forces, attraction of charges, association between polar and non-polar groups, hydrogen bridges, covalent bonds.

Plasma lipoproteins transport lipophilic toxicants such as PCBs. The other plasma fractions serve as a transport vehicle too. The affinity of toxicants for plasma proteins suggests their affinity for proteins in tissues and organs during distribution.

Organic acids (lactic, glutaminic, citric) form complexes with some toxicants. Alkaline earths and rare earths, as well as some heavy elements in the form of cations, are complexed also with organic oxy- and amino acids. All these complexes are usually diffusible and easily distributed in tissues and organs.

Physiologically chelating agents in plasma such as transferrin and metal-lothionein compete with organic acids and amino acids for cations to form stable chelates.

Diffusible free ions, some complexes and some free molecules are easily cleared from the blood into tissues and organs. The free fraction of ions and molecules is in a dynamic equilibrium with the bound fraction. The concentration of a toxicant in blood will govern the rate of its distribution into tissues and organs, or its mobilization from them into the blood.

Distribution of Toxicants in the Organismn

The human organism can be divided into the following compartments. (1) internal organs, (2) skin and muscles, (3) adipose tissues, (4) connective tissue and bones. This classification is mostly based on the degree of vascular (blood) perfusion in a decreasing order. For example internal organs (including the brain), which represent only 12% of the total body weight, receive about 75% of the total blood volume. On the other hand, connective tissues and bones (15% of total body weight) receive only one per cent of the total blood volume.

The well-perfused internal organs generally achieve the highest concentration of toxicants in the shortest time, as well as an equilibrium between blood and this compartment. The uptake of toxicants by less perfused tissues is much slower, but retention is higher and duration of stay much longer (accumulation) due to low perfusion.

Three components are of major importance for the intracellular distribution of toxicants: content of water, lipids and proteins in the cells of various tissues and organs. The above-mentioned order of compartments also follows closely a decreasing water content in their cells. Hydrophilic toxicants will be more rapidly distributed to the body fluids and cells with high water content, and lipophilic toxicants to cells with higher lipid content (fatty tissue).

The organism possesses some barriers which impair penetration of some groups of toxicants, mostly hydrophilic, to certain organs and tissues, such as:

- the blood-brain barrier (cerebrospinal barrier), which restricts penetration of large molecules and hydrophilic toxicants to the brain and CNS; this barrier consists of a closely joined layer of endothelial cells; thus, lipophilic toxicants can penetrate through it

- the placental barrier, which has a similar effect on penetration of toxicants into the foetus from the blood of the mother

- the histo-haematologic barrier in the walls of capillaries, which is permeable for small- and intermediate-sized molecules, and for some larger molecules, as well as ions.

As previously noted only the free forms of toxicants in plasma (molecules, ions, colloids) are available for penetration through the capillary walls participating in distribution. This free fraction is in a dynamic equilibrium with the bound fraction. Concentration of toxicants in blood is in a dynamic equilibrium with their concentration in organs and tissues, governing retention (accumulation) or mobilization from them.

The condition of the organism, functional state of organs (especially neuro-humoral regulation), hormonal balance and other factors play a role in distribution.

Retention of toxicant in a particular compartment is generally temporary and redistribution into other tissues can occur. Retention and accumulation is based on the difference between the rates of absorption and elimination. The duration of retention in a compartment is expressed by the biological half-life. This is the time interval in which 50% of the toxicant is cleared from the tissue or organ and redistributed, translocated or eliminated from the organism.

Biotransformation processes occur during distribution and retention in various organs and tissues. Biotransformation produces more polar, more hydrophilic metabolites, which are more easily eliminated. A low rate of biotransformation of a lipophilic toxicant will generally cause its accumulation in a compartment.

The toxicants can be divided into four main groups according to their affinity, predominant retention and accumulation in a particular compartment:

1. Toxicants soluble in the body fluids are uniformly distributed according to the water content of compartments. Many monovalent cations (*e.g.*, lithium, sodium, potassium, rubidium) and some anions (*e.g.*, chlorine, bromine), are distributed according to this pattern.

2. Lipophilic toxicants show a high affinity for lipid-rich organs (CNS) and tissues (fatty, adipose).

3. Toxicants forming colloid particles are then trapped by specialized cells of the reticuloendothelial system (RES) of organs and tissues. Tri- and quadrivalent cations (lanthanum, cesium, hafnium) are distributed in the RES of tissues and organs.

4. Toxicants showing a high affinity for bones and connective tissue (osteotropic elements, bone seekers) include divalent cations (*e.g.*, calcium, barium, strontium, radon, beryllium, aluminium, cadmium, lead).

Accumulation in Lipid-rich Tissues

The "standard man" of 70 kg body weight contains about 15% of body weight in the form of adipose tissue, increasing with obesity to 50%. However, this lipid fraction is not uniformly distributed. The brain (CNS) is a lipid-rich organ, and peripheral nerves are wrapped with a lipid-rich myelin sheath and Schwann cells. All these tissues offer possibilities for accumulation of lipophilic toxicants.

Numerous non-electrolytes and non-polar toxicants with a suitable Nernst partition coefficient will be distributed to this compartment, as well as numerous organic solvents (alcohols, aldehydes, ketones, *etc.*), chlorinated hydrocarbons (including organochlorine insecticides such as DDT), some inert gases (radon), *etc.*

Adipose tissue will accumulate toxicants due to its low vascularization and lower rate of biotransformation. Here accumulation of toxicants may represent a kind of temporary "neutralization" because of lack of targets for toxic effect. However, potential danger for the organism is always present due to the possibility of mobilization of toxicants from this compartment back to the circulation.

Deposition of toxicants in the brain (CNS) or lipid-rich tissue of the myelin sheath of the peripheral nervous system is very dangerous. The neurotoxicants are deposited here directly next to their targets. Toxicants retained in lipid-rich tissue of the endocrine glands can produce hormonal disturbances. Despite the blood-brain barrier, numerous neurotoxicants of a lipophilic nature reach the brain (CNS): anaesthetics, organic solvents, pesticides, tetraethyl lead, organomercurials, *etc.*

Retention in the Reticuloendothelial System

In each tissue and organ a certain percentage of cells is specialized for phagocytic activity, engulfing micro-organisms, particles, colloid particles, and so on. This system is called the reticuloendothelial system (RES), comprising fixed cells as well as moving cells (phagocytes). These cells are present in non-active form. An increase of the above-mentioned microbes and particles will activate the cells up to a saturation point.

Toxicants in the form of colloids will be captured by the RES of organs and tissues. Distribution depends on the colloid particle size. For larger particles, retention in the liver will be favoured. With smaller colloid particles, more or less uniform distribution will occur between the spleen, bone marrow and liver. Clearance of colloids from the RES is very slow, although small particles are cleared relatively more quickly.

Accumulation in Bones

About 60 elements can be identified as osteotropic elements, or bone seekers.

Osteotropic elements can be divided into three groups:

1. Elements representing or replacing physiological constituents of the bone. Twenty such elements are present in higher quantities. The others appear in trace quantities. Under conditions of chronic exposure, toxic metals such as lead, aluminium and mercury can also enter the mineral matrix of bone cells.

2. Alkaline earths and other elements forming cations with an ionic diameter similar to that of calcium are exchangeable with it in bone mineral. Also, some anions are exchangeable with anions (phosphate, hydroxyl) of bone mineral.

3. Elements forming microcolloids (rare earths) may be adsorbed on the surface of bone mineral.

The skeleton of a standard man accounts for 10 to 15% of the total body weight, representing a large potential storage depot for osteotropic toxicants. Bone is a highly specialized tissue consisting by volume of 54% minerals and 38% organic matrix. The mineral matrix of bone is hydroxyapatite, $Ca_{10}(PO_4)_6(OH)_2$, in which the ratio of Ca to P is about 1.5 to one. The surface area of mineral available for adsorption is about 100 m^2 per g of bone.

Metabolic activity of the bones of the skeleton can be divided in two categories:

*   active, metabolic bone, in which processes of resorption and new bone formation, or remodelling of existing bone, are very extensive

*   stable bone with a low rate of remodelling or growth.

In the fetus, infant and young child metabolic bone (see "available skeleton") represents almost 100% of the skeleton. With age this percentage of metabolic bone decreases. Incorporation of toxicants during exposure appears in the metabolic bone and in more slowly turning-over compartments.

Incorporation of toxicants into bone occurs in two ways:

1. For ions, an ion exchange occurs with physiologically present calcium cations, or anions (phosphate, hydroxyl).

2. For toxicants forming colloid particles, adsorption on the mineral surface occurs.

Ion-exchange Reactions

The bone mineral, hydroxyapatite, represents a complex ion-exchange system. Calcium cations can be exchanged by various cations. The anions present in bone can also be exchanged by anions: phosphate with citrates and carbonates, hydroxyl with fluorine. Ions which are not exchangeable can be adsorbed on the mineral surface. When toxicant ions are incorporated in the mineral, a new layer of mineral can cover the mineral surface, burying toxicant into the bone structure. Ion exchange is a reversible process, depending on the concentration of ions, pH and fluid volume. Thus, for example, an increase of dietary calcium may decrease the deposition of toxicant ions in the lattice of minerals. It has been mentioned that with age the percentage of metabolic bone is decreased, although ion exchange continues. With ageing, bone mineral resorption occurs, in which bone density actually decreases. At this point, toxicants in bone may be released (*e.g.*, lead).

About 30% of the ions incorporated into bone minerals are loosely bound and can be exchanged, captured by natural chelating agents and excreted, with a biological half-life of 15 days. The other 70% is more firmly bound. Mobilization and excretion of this fraction shows a biological half-life of 2.5 years and more depending on bone type (remodelling processes).

Chelating agents (Ca-EDTA, penicillamine, BAL, *etc.*) can mobilize consider-able quantities of some heavy metals, and their excretion in urine greatly increased.

Colloid Adsorption

Colloid particles are adsorbed as a film on the mineral surface (100 m^2 per g) by Van der Waals forces or chemisorption. This layer of colloids on the mineral surfaces is covered with the next layer of formed minerals, and the toxicants are more buried into the bone structure. The rate of mobilization and elimination depends on remodelling processes.

Accumulation in Hair and Nails

The hair and nails contain keratin, with sulphydryl groups able to chelate metallic cations such as mercury and lead.

Distribution of Toxicant Inside the Cell

Recently the distribution of toxicants, especially some heavy metals, within cells of tissues and organs has become of importance. With ultracentrifugation techniques, various fractions of the cell can be separated to determine their content of metal ions and other toxicants.

Animal studies have revealed that after penetration into the cell, some metal ions are bound to a specific protein, metallothionein. This low molecular weight protein is present in the cells of liver, kidney and other organs and tissues. Its sulphydryl groups can bind six ions per molecule. Increased presence of metal ions induces the biosynthesis of this protein. Ions of cadmium are the most potent inducer. Metallothionein serves also to maintain homeostasis of vital copper and zinc ions. Metallothionein can bind zinc, copper, cadmium, mercury, bismuth, gold, cobalt and other cations.

Biotransformation and Elimination of Toxicants

During retention in cells of various tissues and organs, toxicants are exposed to enzymes which can biotransform (metabolize) them, producing metabolites. There are many pathways for the elimination of toxicants and/or metabolites: by exhaled air *via* the lungs, by urine *via* the kidneys, by bile *via* the GIT, by sweat *via* the skin, by saliva *via* the mouth mucosa, by milk *via* the mammary glands, and by hair and nails *via* normal growth and cell turnover.

The elimination of an absorbed toxicant depends on the portal of entry. In the lungs the absorption/desorption process starts immediately and toxicants are partially eliminated by exhaled air. Elimination of toxicants absorbed by other paths of entry is prolonged and starts after transport by blood, eventually being completed after distribution and biotransformation. During absorption an equilibrium exists between the concentrations of a toxicant in the blood and in tissues and organs. Excretion decreases toxicant blood concentration and may induce mobilization of a toxicant from tissues into blood.

Many factors can influence the elimination rate of toxicants and their metabolites from the body:

- physico-chemical properties of toxicants, especially the Nernst partition coefficient (P), dissociation constant (pKa), polarity, molecular structure, shape and weight
- level of exposure and time of post-exposure elimination
- portal of entry
- distribution in the body compartments, which differ in exchange rate with the blood and blood perfusion
- rate of biotransformation of lipophilic toxicants to more hydrophilic metabolites
- overall health condition of organism and, especially, of excretory organs (lungs, kidneys, GIT, skin, *etc.*)
- presence of other toxicants which can interfere with elimination.

Here we distinguish two groups of compartments: (1) the rapid-exchange system— in these compartments, tissue concentration of toxicant is similar to that of the blood; and (2) the slow-exchange system, where tissue concentration of toxicant is higher than in blood due to binding and accumulation—adipose tissue, skeleton and kidneys can temporarily retain some toxicants, *e.g.*, arsenic and zinc.

A toxicant can be excreted simultaneously by two or more excretion routes. However, usually one route is dominant.

Scientists are developing mathematical models describing the excretion of a particular toxicant. These models are based on the movement from one or both compartments (exchange systems), biotransformation and so on.

Elimination by Exhaled Air *via* Lungs

Elimination *via* the lungs (desorption) is typical for toxicants with high volatility (*e.g.*, organic solvents). Gases and vapours with low solubility in blood will be quickly eliminated this way, whereas toxicants with high blood solubility will be eliminated by other routes.

Organic solvents absorbed by the GIT or skin are excreted partially by exhaled air in each passage of blood through the lungs, if they have a sufficient vapour pressure. The Breathalyser test used for suspected drunk drivers is based on this fact. The concentration of CO in exhaled air is in equilibrium with the CO-Hb blood content. The radioactive gas radon appears in exhaled air due to the decay of radium accumulated in the skeleton.

Elimination of a toxicant by exhaled air in relation to the post-exposure period of time usually is expressed by a three-phase curve. The first phase represents elimination of toxicant from the blood, showing a short half-life. The second, slower phase represents elimination due to exchange of blood with tissues and organs (quick-exchange system). The third, very slow phase is due to exchange of blood with fatty tissue and skeleton. If a toxicant is not accumulated in such compartments, the curve will be two-phase. In some cases a four-phase curve is also possible.

Determination of gases and vapours in exhaled air in the post-exposure period is sometimes used for evaluation of exposures in workers.

Renal Excretion

The kidney is an organ specialized in the excretion of numerous water-soluble toxicants and metabolites, maintaining homeostasis of the organism. Each kidney possesses about one million nephrons able to perform excretion. Renal excretion represents a very complex event encompassing three different mechanisms:

- glomerular filtration by Bowman's capsule
- active transport in the proximal tubule
- passive transport in the distal tubule.

Excretion of a toxicant *via* the kidneys to urine depends on the Nernst partition coefficient, dissociation constant and pH of urine, molecular size and shape, rate of metabolism to more hydrophilic metabolites, as well as health status of the kidneys.

The kinetics of renal excretion of a toxicant or its metabolite can be expressed by a two-, three- or four-phase excretion curve, depending on the distribution of the particular toxicant in various body compartments differing in the rate of exchange with the blood.

Saliva

Some drugs and metallic ions can be excreted through the mucosa of the mouth by saliva—for example, lead ("lead line"), mercury, arsenic, copper, as well as bromides, iodides, ethyl alcohol, alkaloids, and so on. The toxicants are then swallowed, reaching the GIT, where they can be reabsorbed or eliminated by faeces.

Sweat

Many non-electrolytes can be partially eliminated *via* skin by sweat: ethyl alcohol, acetone, phenols, carbon disulphide and chlorinated hydrocarbons.

Milk

Many metals, organic solvents and some organochlorine pesticides (DDT) are secreted *via* the mammary gland in mother's milk. This pathway can represent a danger for nursing infants.

Hair

Analysis of hair can be used as an indicator of homeostasis of some physiological substances. Also exposure to some toxicants, especially heavy metals, can be evaluated by this kind of bioassay.

Elimination of toxicants from the body can be increased by:

• mechanical translocation *via* gastric lavage, blood transfusion or dialysis

• creating physiological conditions which mobilize toxicants by diet, change of hormonal balance, improving renal function by application of diuretics

• administration of complexing agents (citrates, oxalates, salicilates, phosphates), or chelating agents (Ca-EDTA, BAL, ATA, DMSA, penicillamine); this method is indicated only in persons under strict medical control. Application of chelating agents is often used for elimination of heavy metals from the body of exposed workers in the course of their medical treatment. This method is also used for evaluation of total body burden and level of past exposure.

Exposure Determinations

Determination of toxicants and metabolites in blood, exhaled air, urine, sweat, faeces and hair is more and more used for evaluation of human exposure (exposure tests) and/or evaluation of the degree of intoxication. Therefore biological exposure limits (Biological MAC Values, Biological Exposure Indices — BEI) have recently been established. These bioassays show "internal exposure" of the organism, that is, total exposure of the body in both the work and living environments by all portals of entry (see "Toxicology test methods: Biomarkers").

Combined Effects Due to Multiple Exposure

People in the work and/or living environment are usually exposed simultaneously or consecutively to various physical and chemical agents. Also it is necessary to take into consideration that some persons use medications, smoke, consume alcohol and food containing additives and so on. That means that usually multiple exposure is occurring. Physical and chemical agents can interact in each step of toxicokinetic and/or toxicodynamic processes, producing three possible effects:

1. Independent. Each agent produces a different effect due to a different mechanism of action.

2. Synergistic. The combined effect is greater than that of each single agent. Here we differentiate two types: (a) additive, where the combined effect

is equal to the sum of the effects produced by each agent separately and (b) potentiating, where the combined effect is greater than additive.

3. Antagonistic. The combined effect is lower than additive.

However, studies on combined effects are rare. This kind of study is very complex due to the combination of various factors and agents.

We can conclude that when the human organism is exposed to two or more toxicants simultaneously or consecutively, it is necessary to consider the possibility of some combined effects, which can increase or decrease the rate of toxicokinetic processes.

TARGET ORGAN AND CRITICAL EFFECTS

The priority objective of occupational and environmental toxicology is to improve the prevention or substantial limitation of health effects of exposure to hazardous agents in the general and occupational environments. To this end systems have been developed for quantitative risk assessment related to a given exposure (see the section "Regulatory toxicology").

The effects of a chemical on particular systems and organs are related to the magnitude of exposure and whether exposure is acute or chronic. In view of the diversity of toxic effects even within one system or organ, a uniform philosophy concerning the critical organ and critical effect has been proposed for the purpose of risk assessment and development of health-based recommended concentration limits of toxic substances in different environmental media.

From the point of view of preventive medicine, it is of particular importance to identify early adverse effects, based on the general assumption that preventing or limiting early effects may prevent more severe health effects from developing.

Such an approach has been applied to heavy metals. Although heavy metals, such as lead, cadmium and mercury, belong to a specific group of toxic substances where the chronic effect of activity is dependent on their accumulation in the organs, the definitions presented below were published by the Task Group on Metal Toxicity (Nordberg 1976).

The definition of the critical organ as proposed by the Task Group on Metal Toxicity has been adopted with a slight modification: the word metal has been replaced with the expression potentially toxic substance (Duffus 1993).

Whether a given organ or system is regarded as critical depends not only on the toxicomechanics of the hazardous agent but also on the route of absorption and the exposed population.

- Critical concentration for a cell: the concentration at which adverse functional changes, reversible or irreversible, occur in the cell.
- Critical organ concentration: the mean concentration in the organ at the time at which the most sensitive type of cells in the organ reach critical concentration.

- Critical organ: that particular organ which first attains the critical concentration of metal under specified circumstances of exposure and for a given population.
- Critical effect: defined point in the relationship between dose and effect in the individual, namely the point at which an adverse effect occurs in cellular function of the critical organ. At an exposure level lower than that giving a critical concentration of metal in the critical organ, some effects may occur that do not impair cellular function per se, yet are detectable by means of biochemical and other tests. Such effects are defined as subcritical effects.

The biological meaning of subcritical effect is sometimes not known; it may stand for exposure biomarker, adaptation index or a critical effect precursor (see "Toxicology test methods: Biomarkers"). The latter possibility can be particularly significant in view of prophylactic activities.

In chronic environmental exposure to cadmium, where the route of absorption is of minor importance (cadmium air concentrations range from 10 to 20 µg/m3 in the urban and 1 to 2 µg/m3 in the rural areas), the critical organ is the kidney. In the occupational setting where the TLV reaches 50 µg/m3 and inhalation constitutes the main route of exposure, two organs, lung and kidney, are regarded as critical.

Table. Examples of critical organs and critical effects

Substance	Critical organ in chronic exposure	Critical effect
Cadmium	Lungs	Nonthreshold: Lung cancer (unit risk 4.6×10^{-3})
	Kidney	Threshold: Increased excretion of low molecular proteins (β_2 -M, RBP) in urine
	Lungs	Emphysema slight function changes
	Adults	
Lead	Haematopoietic system	Increased delta-aminolevulinic acid excretion in urine (ALA-U); increased concentration of free erythrocyte protoporphyrin (FEP) in erythrocytes
	Peripheral nervous system	Slowing of the conduction velocities of the slower nerve fibres
	Young children	
Mercury (elemental)	Central nervous system	Decrease in IQ and other subtle effects; mercurial tremor (fingers, lips, eyelids)
Mercury (mercuric)	Kidney	Proteinuria

	Adults	
Manganese	Central nervous system	Impairment of psychomotor functions
	Children	
	Lungs	Respiratory symptoms
	Central nervous system	Impairment of psychomotor functions
Toluene	Mucous membranes	Irritation
Vinyl chloride	Liver	Cancer (angiosarcoma unit risk 1 x 10^{-6})
Ethyl acetate	Mucous membrane	Irritation

For lead, the critical organs in adults are the haemopoietic and peripheral nervous systems, where the critical effects (*e.g.*, elevated free erythrocyte protoporphyrin concentration (FEP), increased excretion of delta-aminolevulinic acid in urine, or impaired peripheral nerve conduction) manifest when the blood lead level (an index of lead absorption in the system) approaches 200 to 300 µg/l. In small children the critical organ is the central nervous system (CNS), and the symptoms of dysfunction detected with the use of a psychological test battery have been found to appear in the examined populations even at concentrations in the range of about 100 µg/l Pb in blood.

A number of other definitions have been formulated which may better reflect the meaning of the notion. According to WHO (1989), the critical effect has been defined as "the first adverse effect which appears when the threshold (critical) concentration or dose is reached in the critical organ. Adverse effects, such as cancer, with no defined threshold concentration are often regarded as critical. Decision on whether an effect is critical is a matter of expert judgement." In the International Programme on Chemical Safety (IPCS) guidelines for developing Environmental Health Criteria Documents, the critical effect is described as "the adverse effect judged to be most appropriate for determining the tolerable intake". The latter definition has been formulated directly for the purpose of evaluating the health-based exposure limits in the general environment. In this context the most essential seems to be determining which effect can be regarded as an adverse effect. Following current terminology, the adverse effect is the "change in morphology, physiology, growth, development or lifespan of an organism which results in impairment of the capacity to compensate for additional stress or increase in susceptibility to the harmful effects of other environmental influences. Decision on whether or not any effect is adverse requires expert judgement."

In the case of exposure to lead, A can represent a subcritical effect (inhibition of erythrocyte ALA-dehydratase), B the critical effect (an increase in erythrocyte zinc protoporphyrin or increase in the excretion of delta-aminolevulinic acid, C the clinical effect (anaemia) and D the fatal effect (death). For lead exposure there is abundant evidence illustrating how particular effects of exposure are dependent on lead concentration in blood (practical counterpart of the dose), either in the form of the dose-response relationship or in relation to different variables (sex, age, *etc.*). Determining the critical effects and the dose-response relationship for such effects in humans makes it possible to predict the frequency of a given effect for a given dose or its counterpart (concentration in biological material) in a certain population.

The critical effects can be of two types: those considered to have a threshold and those for which there may be some risk at any exposure level (non-threshold, genotoxic carcinogens and germ mutagens). Whenever possible, appropriate human data should be used as a basis for the risk assessment. In order to determine the threshold effects for the general population, assumptions concerning the exposure level (tolerable intake, biomarkers of exposure) have to be made such that the frequency of the critical effect in the population exposed to a given hazardous agent corresponds to the frequency of that effect in the general population. In lead exposure, the maximum recommended blood lead concentration for the general population (200 µg/l, median below 100 µg/l) (WHO 1987) is practically below the threshold value for the assumed critical effect-the elevated free erythrocyte protoporphyrin level, although it is not below the level associated with effects on the CNS in children or blood pressure in adults. In general, if data from well-conducted human population studies defining a no observed adverse effect level are the basis for safety evaluation, then the uncertainty factor of ten has been considered appropriate. In the case of occupational exposure the critical effects may refer to a certain part of the population (*e.g.* 10%). Accordingly, in occupational lead exposure the recommended health-based concentration of blood lead has been adopted to be 400 mg/l in men where a 10% response level for ALA-U of 5 mg/l occurred at PbB concentrations of about 300 to 400 mg/l. For the occupational exposure to cadmium (assuming the increased urinary excretion of low-weight proteins to be the critical effect), the level of 200 ppm cadmium in renal cortex has been regarded as the admissible value, for this effect has been observed in 10% of the exposed population. Both these values are under consideration for lowering, in many countries, at the present time (*i.e.*,1996).

There is no clear consensus on appropriate methodology for the risk assessment of chemicals for which the critical effect may not have a threshold, such as genotoxic carcinogens. A number of approaches based largely on characterization of the dose- response relationship have been adopted for the assessment of such effects. Owing to the lack of socio-political acceptance of health risk caused by carcinogens in such documents as the Air Quality Guidelines for Europe (WHO 1987), only the values such as the unit lifetime risk (*i.e.*, the risk associated with lifetime exposure to 1 µg/m3 of the hazardous agent) are presented for non threshold effects (see "Regulatory toxicology").

Presently, the basic step in undertaking activities for risk assessment is determining the critical organ and critical effects. The definitions of both the critical and adverse effect reflect the responsibility of deciding which of the effects within a given organ or system should be regarded as critical, and this is directly related to the subsequent determination of recommended values for a given chemical in the general environment-for example, Air Quality Guidelines for Europe (WHO 1987) or health-based limits in occupational exposure (WHO 1980). Determining the critical effect from within the range of subcritical effects may lead to a situation where the recommended limits on toxic chemicals concentration in the general or occupational environment may be in practice impossible to maintain. Regarding as critical an effect that may overlap the early clinical effects may bring about the adoption of the values for which adverse effects may develop in some part of the population. The decision whether or not a given effect should be considered critical remains the responsibility of expert groups who specialize in toxicity and risk assessment.

EFFECTS OF AGE, SEX AND OTHER FACTORS

There are often large differences among humans in the intensity of response to toxic chemicals, and variations in susceptibility of an individual over a lifetime. These can be attributed to a variety of factors capable of influencing absorption rate, distribution in the body, biotransformation and/or excretion rate of a particular chemical. Apart from the known hereditary factors which have been clearly demonstrated to be linked with increased susceptibility to chemical toxicity in humans (see "Genetic determinants of toxic response"), other factors include: constitutional characteristics related to age and sex; pre-existing disease states or a reduction in organ function (non-hereditary, *i.e.*, acquired); dietary habits, smoking, alcohol consumption and use of medications; concomitant exposure to biotoxins (various microorganisms) and physical factors (radiation, humidity, extremely low or high temperatures or barometric pressures particularly relevant to the partial pressure of a gas), as well as concomitant physical exercise or psychological stress situations; previous occupational and/or environmental exposure to a particular chemical, and in particular concomitant exposure to other chemicals, not necessarily toxic (*e.g.*, essential metals). The possible contributions of the aforementioned factors in either increasing or decreasing susceptibility to adverse health effects, as well as the mechanisms of their action, are specific for a particular chemical. Therefore only the most common factors, basic mechanisms and a few characteristic examples will be presented here, whereas specific information concerning each particular chemical can be found in elsewhere in this Encyclopaedia.

According to the stage at which these factors act (absorption, distribution, biotransformation or excretion of a particular chemical), the mechanisms can be roughly categorized according to two basic consequences of interaction: (1) a change in the quantity of the chemical in a target organ, that is, at the site(s) of its effect in the organism (toxicokinetic interactions), or (2) a change in the intensity of a specific response to the quantity of the chemical in a target organ (toxicodynamic interactions). The most common mechanisms of either type of interaction are related to competition with other chemical(s) for binding to the same compounds involved in their transport in the organism (*e.g.*, specific serum proteins) and/or for the same biotransformation pathway (*e.g.*, specific enzymes) resulting in a change in the speed or sequence between initial reaction and final adverse health effect. However, both toxicokinetic and toxicodynamic interactions may influence individual susceptibility to a particular chemical. The influence of several concomitant factors can result in either: (a) additive effects – the intensity of the combined effect is equal to the sum of the effects produced by each factor separately, (b) synergistic effects – the intensity of the combined effect is greater than the sum of the effects produced by each factor separately, or (c) antagonistic effects – the intensity of the combined effect is smaller than the sum of the effects produced by each factor separately.

The quantity of a particular toxic chemical or characteristic metabolite at the site(s) of its effect in the human body can be more or less assessed by biological monitoring, that is, by choosing the correct biological specimen and optimal timing of specimen sampling, taking into account biological half-lives for a particular chemical in both the critical organ and in the measured biological compartment. However, reliable information concerning other possible factors that might influence individual susceptibility in humans is generally lacking, and consequently the majority of knowledge regarding the influence of various factors is based on experimental animal data.

It should be stressed that in some cases relatively large differences exist between humans and other mammals in the intensity of response to an equivalent level and/or duration of exposure to many toxic chemicals; for example, humans appear to be considerably more sensitive to the adverse health effects of several toxic metals than are rats (commonly used in experimental animal studies). Some of these differences can be attributed to the fact that the transportation, distribution and biotransformation pathways of various chemicals are greatly dependent on subtle changes in the tissue pH and the redox equilibrium in the organism (as are the activities of various enzymes), and that the redox system of the human differs considerably from that of the rat.

This is obviously the case regarding important antioxidants such as vitamin C and glutathione (GSH), which are essential for maintaining redox equilibrium and which have a protective role against the adverse effects of the oxygen- or xenobiotic-derived free radicals which are involved in a variety of pathological conditions. Humans cannot auto-synthesize vitamin C, contrary to the rat, and levels as well as the turnover rate of erythrocyte GSH in humans are considerably lower than that in the rat. Humans also lack some of the protective antioxidant enzymes, compared to the rat or other mammals (*e.g.*, GSH- peroxidase is considered to be poorly active in human sperm). These examples illustrate the potentially greater vulnerability to oxidative stress in humans (particularly in sensitive cells, *e.g.*, apparently greater vulnerability of the human sperm to toxic influences than that of the rat), which can result in different response or greater susceptibility to the influence of various factors in humans compared to other mammals.

Influence of Age

Compared to adults, very young children are often more susceptible to chemical toxicity because of their relatively greater inhalation volumes and gastrointestinal absorption rate due to greater permeability of the intestinal epithelium, and because of immature detoxification enzyme systems and a relatively smaller excretion rate of toxic chemicals. The central nervous system appears to be particularly susceptible at the early stage of development with regard to neurotoxicity of various chemicals, for example, lead and methylmercury. On the other hand, the elderly may be susceptible because of chemical exposure history and increased body stores of some xenobiotics, or pre-existing compromised function of target organs and/or relevant enzymes resulting in lowered detoxification and excretion rate. Each of these factors can contribute to weakening of the body's defences — a decrease in reserve capacity, causing increased susceptibility to subsequent exposure to other hazards. For example, the cytochrome P450 enzymes (involved in the biotransformation pathways of almost all toxic chemicals) can be either induced or have lowered activity because of the influence of various factors over a lifetime (including dietary habits, smoking, alcohol, use of medications and exposure to environmental xenobiotics).

Influence of Sex

Gender-related differences in susceptibility have been described for a large number of toxic chemicals (approximately 200), and such differences are found in many mammalian species. It appears that males are generally more susceptible to renal toxins and females to liver toxins. The causes of the different response between males and females have been related to differences in a variety of physiological processes (*e.g.*, females are capable of additional excretion of some toxic chemicals through menstrual blood loss, breast milk and/or transfer to the foetus, but they experience additional stress during pregnancy, delivery and lactation), enzyme activities, genetic repair mechanisms, hormonal factors, or the presence of relatively larger fat depots in females, resulting in greater accumulation of some lipophilic toxic chemicals, such as organic solvents and some medications.

Influence of Dietary Habits

Dietary habits have an important influence on susceptibility to chemical toxicity, mostly because adequate nutrition is essential for the functioning of the body's chemical defence system in maintaining good health. Adequate intake of essential metals (including metalloids) and proteins, especially the sulphur-containing amino acids, is necessary for the biosynthesis of various detoxificating enzymes and the provision of glycine and glutathione for conjugation reactions with endogenous and exogenous compounds. Lipids, especially phospholipids, and lipotropes (methyl group donors) are necessary for the synthesis of biological membranes. Carbohydrates provide the energy required for various detoxification processes and provide glucuronic acid for conjugation of toxic chemicals and their metabolites. Selenium (an essential metalloid), glutathione, and vitamins such as vitamin C (water soluble), vitamin E and vitamin A (lipid soluble), have an important role as antioxidants (*e.g.*, in controlling lipid peroxidation and maintaining integrity of cellular membranes) and free-radical scavengers for protection against toxic chemicals.

In addition, various dietary constituents (protein and fibre content, minerals, phosphates, citric acid, *etc.*) as well as the amount of food consumed can greatly influence the gastrointestinal absorption rate of many toxic chemicals (*e.g.*, the average absorption rate of soluble lead salts taken with meals is approximately eight per cent, as opposed to approximately 60% in fasting subjects). However, diet itself can be an additional source of individual exposure to various toxic chemicals (*e.g.*, considerably increased daily intakes and accumulation of arsenic, mercury, cadmium and/or lead in subjects who consume contaminated seafood).

Influence of Smoking

The habit of smoking can influence individual susceptibility to many toxic chemicals because of the variety of possible interactions involving the great number of compounds present in cigarette smoke (especially polycyclic aromatic hydrocarbons, carbon monoxide, benzene, nicotine, acrolein, some pesticides, cadmium, and, to a lesser extent, lead and other toxic metals, *etc.*), some of which are capable of accumulating in the human body over a lifetime, including pre-natal life (*e.g.*, lead and cadmium). The interactions occur mainly because various toxic chemicals compete for the same binding site(s) for transport and distribution in the organism and/or for the same biotransformation pathway involving particular enzymes. For example, several cigarette smoke constituents can induce cytochrome P450 enzymes, whereas others can depress their activity, and thus influence the common biotransformation pathways of many other toxic chemicals, such as organic solvents and some medications. Heavy cigarette smoking over a long period can considerably reduce the body's defence mechanisms by decreasing reserve capacity to cope with the adverse influence of other life-style factors.

Influence of Alcohol

Consumption of alcohol (ethanol) can influence susceptibility to many toxic chemicals in several ways. It can influence the absorption rate and distribution of certain chemicals in the body — for example, increase the gastrointestinal absorption rate of lead, or decrease the pulmonary absorption rate of mercury vapour by inhibiting oxidation which is necessary for retention of inhaled mercury vapour. Ethanol can also influence susceptibility to various chemicals through short-term changes in tissue pH and increase in the redox potential resulting from ethanol metabolism, as both ethanol oxidizing to acetaldehyde and acetaldehyde oxidizing to acetate produce an equivalent of reduced nicotinamide adenine dinucleotide (NADH) and hydrogen (H+). Because the affinity of both essential and toxic metals and metalloids for binding to various compounds and tissues is influenced by pH and changes in the redox potential, even a moderate intake of ethanol may result in a series of consequences such as: (1) redistribution of long-term accumulated lead in the human organism in favour of a biologically active lead fraction, (2) replacement of essential zinc by lead in zinc-containing enzyme(s), thus affecting enzyme activity, or influence of mobilized lead on the distribution of other essential metals and metalloids in the organism such as calcium, iron, copper and selenium, (3) increased urinary excretion of zinc and so on. The effect of possible aforementioned events can be augmented due to the fact that alcoholic beverages can contain an appreciable amount of lead from vessels or processing.

Another common reason for ethanol-related changes in susceptibility is that many toxic chemicals, for example, various organic solvents, share the same biotransformation pathway involving the cytochrome P450 enzymes. Depending on the intensity of exposure to organic solvents as well as the quantity and frequency of ethanol ingestion (*i.e.*, acute or chronic alcohol consumption), ethanol can either decrease or increase biotransformation rates of various organic solvents and thus influence their toxicity.

Influence of Medications

The common use of various medications can influence susceptibility to toxic chemicals mainly because many drugs bind to serum proteins and thus influence the transport, distribution or excretion rate of various toxic chemicals, or because many drugs are capable of inducing relevant detoxifying enzymes or depressing their activity (*e.g.*, the cytochrome P450 enzymes), thus affecting the toxicity of chemicals with the same biotransformation pathway. Characteristic for either of the mechanisms is increased urinary excretion of trichloroacetic acid (the metabolite of several chlorinated hydrocarbons) when using salicylate, sulphonamide or phenylbutazone, and an increased hepato-nephrotoxicity of carbon tetrachloride when using phenobarbital. In addition, some medications contain a considerable amount of a potentially toxic chemical, for example, the aluminium-containing antacids or preparations used for therapeutic management of the hyperphosphataemia arising in chronic renal failure.

Influence of Concomitant Exposure to Other Chemicals

The changes in susceptibility to adverse health effects due to interaction of various chemicals (*i.e.*, possible additive, synergistic or antagonistic effects) have been studied almost exclusively in experimental animals, mostly in the rat. Relevant epidemiological and clinical studies are lacking. This is of concern particularly considering the relatively greater intensity of response or the variety of adverse health effects of several toxic chemicals in humans compared to the rat and other mammals. Apart from published data in the field of pharmacology, most data are related only to combinations of two different chemicals within specific groups, such as various pesticides, organic solvents, or essential and/or toxic metals and metalloids.

Combined exposure to various organic solvents can result in various additive, synergistic or antagonistic effects (depending on the combination of certain organic solvents, their intensity and duration of exposure), mainly due to the capability of influencing each other's biotransformation.

Another characteristic example are the interactions of both essential and/ or toxic metals and metalloids, as these are involved in the possible influence of age (*e.g.*, a lifetime body accumulation of environmental lead and cadmium), sex (*e.g.*, common iron deficiency in women), dietary habits (*e.g.*, increased dietary intake of toxic metals and metalloids and/or deficient dietary intake of essential metals and metalloids), smoking habit and alcohol consumption (*e.g.*, additional exposure to cadmium, lead and other toxic metals), and use of medications (*e.g.*, a single dose of antacid can result in a 50-fold increase in the average daily intake of aluminium through food). The possibility of various additive, synergistic or antagonistic effects of exposure to various metals and metalloids in humans can be illustrated by basic examples related to the main toxic elements, apart from which further interactions may occur because essential elements can also influence one another (*e.g.*, the well-known antagonistic effect of copper on the gastrointestinal absorption rate as well as the metabolism of zinc, and vice versa). The main cause of all these interactions is the competition of various metals and metalloids for the same binding site (especially the sulphhydryl group, -SH) in various enzymes, metalloproteins (especially metallothionein) and tissues (*e.g.*, cell membranes and organ barriers). These interactions may have a relevant role in the development of several chronic diseases which are mediated through the action of free radicals and oxidative stress.

Table. Basic effects of possible multiple interactions concerning the main toxic and/or essential metals and matalloids in mammals

Toxic metal or metalloid	Basic effects of the interaction with other metal or metalloid
Aluminium (Al)	Decreases the absorption rate of Ca and impairs the metabolism of Ca; deficient dietary Ca increases the absorption rate of Al. Impairs phosphate metabolism. Data on interactions with Fe, Zn and Cu are equivocal (*i.e.*, the possible role of another metal as a mediator).
Arsenic (As)	Affects the distribution of Cu (an increase of Cu in the kidney, and a decrease of Cu in the liver, serum and urine). Impairs the metabolism of Fe (an increase of Fe in the liver with con-comitant decrease in haematocrit). Zn decreases the absorption rate of inorganic As and decreases the tox-icity of As. Se decreases the toxicity of As and vice versa.
Cadmium (Cd)	Decreases the absorption rate of Ca and impairs the metabolism of Ca; deficient dietary Ca increases the absorption rate of Cd. Impairs the phosphate metabolism, *i.e.*, increases urinary excretion of phosphates. Impairs the metabolism of Fe; deficient dietary Fe increases the absorp-tion rate of Cd. Affects the distribution of Zn; Zn decreases the toxicity of Cd, whereas its influence on the absorption rate of Cd is equivocal. Se decreases the toxicity of Cd. Mn decreases the toxicity of Cd at low-level exposure to Cd. Data on the interaction with Cu are equivocal (*i.e.*, the possible role of Zn, or another metal, as a mediator). High dietary levels of Pb, Ni, Sr, Mg or Cr(III) can decrease the absorp-tion rate of Cd.

Mercury (Hg)	Affects the distribution of Cu (an increase of Cu in the liver). Zn decreases the absorption rate of inorganic Hg and decreases the toxicity of Hg. Se decreases the toxicity of Hg. Cd increases the concentration of Hg in the kidney, but at the same time decreases the toxicity of Hg in the kidney (the influence ofthe Cd-induced metallothionein synthesis).
Lead (Pb)	Impairs the metabolism of Ca; deficient dietary Ca increases the absorption rate of inorganic Pb and increases the toxicity of Pb. Impairs the metabolism of Fe; deficient dietary Fe increases the toxicity of Pb, whereas its influence on the absorption rate of Pb is equivocal. Impairs the metabolism of Zn and increases urinary excretion of Zn; deficient dietary Zn increases the absorption rate of inorganic Pb and increases the toxicity of Pb. Se decreases the toxicity of Pb. Data on interactions with Cu and Mg are equivocal (*i.e.*, the possible role of Zn, or another metal, as a mediator).

Note: Data are mostly related to experimental studies in the rat, whereas relevant clinical and epidemiological data (particularly regarding quantitative dose-response relationships) are generally lacking.

GENETIC DETERMINANTS OF TOXIC RESPONSE

It has long been recognized that each person's response to environmental chemicals is different. The recent explosion in molecular biology and genetics has brought a clearer understanding about the molecular basis of such variability. Major determinants of individual response to chemicals include important differences among more than a dozen superfamilies of enzymes, collectively termed xenobiotic- (foreign to the body) or drug-metabolizing enzymes. Although the role of these enzymes has classically been regarded as detoxification, these same enzymes also convert a number of inert compounds to highly toxic intermediates. Recently, many subtle as well as gross differences in the genes encoding these enzymes have been identified, which have been shown to result in marked variations in enzyme activity.

It is now clear that each individual possesses a distinct complement of xenobiotic-metabolizing enzyme activities; this diversity might be thought of as a "metabolic fingerprint". It is the complex interplay of these many different enzyme superfamilies which ultimately determines not only the fate and the potential for toxicity of a chemical in any given individual, but also assessment of exposure. In this article we have chosen to use the cytochrome P450 enzyme superfamily to illustrate the remarkable progress made in understanding individual response to chemicals. The development of relatively simple DNA-based tests designed to identify specific gene alterations in these enzymes, is now providing more accurate predictions of individual response to chemical exposure. We hope the result will be preventive toxicology. In other words, each individual might learn about those chemicals to which he or she is particularly sensitive, thereby avoiding previously unpredictable toxicity or cancer.

Although it is not generally appreciated, human beings are exposed daily to a barrage of innumerable diverse chemicals. Many of these chemicals are highly toxic, and they are derived from a wide variety of environmental and dietary sources. The relationship between such exposures and human health has been, and continues to be, a major focus of biomedical research efforts worldwide.

What are some examples of this chemical bombardment? More than 400 chemicals from red wine have been isolated and characterized. At least 1,000 chemicals are estimated to be produced by a lighted cigarette. There are countless chemicals in cosmetics and perfumed soaps. Another major source of chemical exposure is agriculture: in the United States alone, farmlands receive more than 75,000 chemicals each year in the form of pesticides, herbicides and fertilizing agents; after uptake by plants and grazing animals, as well as fish in nearby waterways, humans (at the end of the food chain) ingest these chemicals. Two other sources of large concentrations of chemicals taken into the body include (a) drugs taken chronically and (b) exposure to hazardous substances in the workplace over a lifetime of employment.

It is now well established that chemical exposure may adversely affect many aspects of human health, causing chronic diseases and the development of many cancers. In the last decade or so, the molecular basis of many of these relationships has begun to be unravelled. In addition, the realization has emerged that humans differ markedly in their susceptibility to the harmful effects of chemical exposure.

Current efforts to predict human response to chemical exposure combine two fundamental approaches (monitoring the extent of human exposure through biological markers (biomarkers), and predicting the likely response of an individual to a given level of exposure. Although both of these approaches are extremely important, it should be emphasized that the two are distinctly different from one another. This article will focus on the genetic factors underlying individual susceptibility to any particular chemical exposure. This field of research is broadly termed ecogenetics, or pharmacogenetics. Many of the recent advances in determining individual susceptibility to chemical toxicity have evolved from a greater appreciation of the processes by which humans and other mammals detoxify chemicals, and the remarkable complexity of the enzyme systems involved.

Toxicologists and pharmacologists commonly speak about the average lethal dose for 50% of the population (LD_{50}), the average maximal tolerated dose for 50% of the population (MTD_{50}), and the average effective dose of a particular drug for 50% of the population (ED_{50}). However, how do these doses affect each of us on an individual basis? In other words, a highly sensitive individual may be 500 times more affected or 500 times more likely to be affected than the most resistant individual in a population; for these people, the LD_{50} (and MTD_{50} and ED_{50}) values would have little meaning. LD_{50}, MTD_{50} and ED_{50} values are only relevant when referring to the population as a whole.

This generic diagram might represent bronchogenic carcinoma in response to the number of cigarettes smokes, chloracne as a function of dioxin levels in the workplace, asthma as a function of air concentrations of ozone or aldehyde, sunburn in response to ultraviolet light, decreased clotting time as a function of aspirin intake, or gastrointestinal distress in response to the number of jalapeño peppers consumed. Generally, in each of these instances, the greater the exposure, the greater the toxic response. Most of the population will exhibit the mean and standard deviation of toxic response as a function of dose. The "resistant outlier" is an individual having less of a response at higher doses or exposures. A "sensitive outlier" (upper left) is an individual having an exaggerated response to a relatively small dose or exposure. These outliers, with extreme differences in response compared to the majority of individuals in the population, may represent important genetic variants that can help scientists in attempting to understand the underlying molecular mechanisms of a toxic response.

Using these outliers in family studies, scientists in a number of laboratories have begun to appreciate the importance of Mendelian inheritance for a given toxic response. Subsequently, one can then turn to molecular biology and genetic studies to pinpoint the underlying mechanism at the gene level (genotype) responsible for the environmentally caused disease (phenotype).

Xenobiotic- or Drug-metabolizing Enzymes

How does the body respond to the myriad of exogenous chemicals to which we are exposed? Humans and other mammals have evolved highly complex metabolic enzyme systems comprising more than a dozen distinct superfamilies of enzymes. Almost every chemical to which humans are exposed will be modified by these enzymes, in order to facilitate removal of the foreign substance from the body. Collectively, these enzymes are frequently referred to as drug-metabolizing enzymes or xenobiotic-metabolizing enzymes. Actually, both terms are misnomers. First, many of these enzymes not only metabolize drugs but hundreds of thousands of environmental and dietary chemicals. Second, all of these enzymes also have normal body compounds as substrates; none of these enzymes metabolizes only foreign chemicals.

For more than four decades, the metabolic processes mediated by these enzymes have commonly been classified as either Phase I or Phase II reactions.

Phase I ("functionalization") reactions generally involve relatively minor structural modifications of the parent chemical *via* oxidation, reduction or hydrolysis in order to produce a more water-soluble metabolite. Frequently, Phase I reactions provide a "handle" for further modification of a compound by subsequent Phase II reactions. Phase I reactions are primarily mediated by a superfamily of highly versatile enzymes, collectively termed cytochromes P450, although other enzyme superfamilies can also be involved

Phase II reactions involve the coupling of a water-soluble endogenous molecule to a chemical (parent chemical or Phase I metabolite) in order to facilitate excretion. Phase II reactions are frequently termed "conjugation" or "derivatization" reactions. The enzyme superfamilies catalyzing Phase II reactions are generally named according to the endogenous conjugating moiety involved: for example, acetylation by the N-acetyltransferases, sulphation by the sulphotransferases, glutathione conjugation by the glutathione transferases, and glucuronidation by the UDP glucuronosyltransferases. Although the major organ of drug metabolism is the liver, the levels of some drug- metabolizing enzymes are quite high in the gastrointestinal tract, gonads, lung, brain and kidney, and such enzymes are undoubtedly present to some extent in every living cell.

Xenobiotic-metabolizing Enzymes Represent Double-edged Swords

As we learn more about the biological and chemical processes leading to human health aberrations, it has become increasingly evident that drug-metabolizing enzymes function in an ambivalent manner. In the majority of cases, lipid-soluble chemicals are converted to more readily excreted water-soluble metabolites. However, it is clear that on many occasions the same enzymes are capable of transforming other inert chemicals into highly reactive molecules. These intermediates can then interact with cellular macromolecules such as proteins and DNA. Thus, for each chemical to which humans are exposed, there exists the potential for the competing pathways of metabolic activation and detoxification.

Brief Review of Genetics

In human genetics, each gene (locus) is located on one of the 23 pairs of chromosomes. The two alleles (one present on each chromosome of the pair) can be the same, or they can be different from one another. For example, the B and b alleles, in which B (brown eyes) is dominant over b (blue eyes): individuals of the brown-eyed phenotype can have either the BB or Bb genotypes, whereas individuals of the blue-eyed phenotype can only have the bb genotype.

A polymorphism is defined as two or more stably inherited phenotypes (traits) — derived from the same gene(s) — that are maintained in the population, often for reasons not necessarily obvious. For a gene to be polymorphic, the gene product must not be essential for development, reproductive vigour or other critical life processes. In fact, a "balanced polymorphism," wherein the heterozygote has a distinct survival advantage over either homozygote (*e.g.*, resistance to malaria, and the sickle-cell haemoglobin allele) is a common explanation for maintaining an allele in the population at otherwise unexplained high.

Human Polymorphisms of Xenobiotic-metabolizing Enzymes

Genetic differences in the metabolism of various drugs and environmental chemicals have been known for more than four decades (Kalow 1962 and 1992). These differences are frequently referred to as pharmacogenetic or, more broadly, ecogenetic polymorphisms. These polymorphisms represent variant alleles that occur at a relatively high frequency in the population and are generally associated with aberrations in enzyme expression or function. Historically, polymorphisms were usually identified following unexpected responses to therapeutic agents. More recently, recombinant DNA technology has enabled scientists to identify the precise alterations in genes that are responsible for some of these polymorphisms. Polymorphisms have now been characterized in many drug-metabolizing enzymes — including both Phase I and Phase II enzymes. As more and more polymorphisms are identified, it is becoming increasingly apparent that each individual may possess a distinct complement of drug-metabolizing enzymes. This diversity might be described as a "metabolic fingerprint". It is the complex interplay of the various drug- metabolizing enzyme superfamilies within any individual that will ultimately determine his or her particular response to a given chemical.

Expressing Human Xenobiotic-metabolizing Enzymes in Cell Culture

How might we develop better predictors of human toxic responses to chemicals? Advances in defining the multiplicity of drug-metabolizing enzymes must be accompanied by precise knowledge as to which enzymes determine the metabolic fate of individual chemicals. Data gleaned from laboratory rodent studies have certainly provided useful information. However, significant interspecies differences in xenobiotic-metabolizing enzymes necessitate caution in extrapolating data to human populations. To overcome this difficulty, many laboratories have developed systems in which various cell lines in culture can be engineered to produce functional human enzymes that are stable and in high concentrations (Gonzalez, Crespi and Gelboin 1991). Successful production of human enzymes has been achieved in a variety of diverse cell lines from sources including bacteria, yeast, insects and mammals.

In order to define the metabolism of chemicals even more accurately, multiple enzymes have also been successfully produced in a single cell line. Such cell lines provide valuable insights into the precise enzymes involved in the metabolic processing of any given compound and likely toxic metabolites. If this information can then be combined with knowledge regarding the presence and level of an enzyme in human tissues, these data should provide valuable predictors of response.

CYTOCHROME P450

History and Nomenclature

The cytochrome P450 superfamily is one of the most studied drug-metabolizing enzyme superfamilies, having a great deal of individual variability in response to chemicals. Cytochrome P450 is a convenient generic term used to describe a large superfamily of enzymes pivotal in the metabolism of innumerable endogenous and exogenous substrates. The term cytochrome P450 was first coined in 1962 to describe an unknown pigment in cells which, when reduced and bound with carbon monoxide, produced a characteristic absorption peak at 450 nm. Since the early 1980s, cDNA cloning technology has resulted in remarkable insights into the multiplicity of cytochrome P450 enzymes. To date, more than 400 distinct cytochrome P450 genes have been identified in animals, plants, bacteria and yeast. It has been estimated that any one mammalian species, such as humans, may possess 60 or more distinct P450 genes. The multiplicity of P450 genes has necessitated the development of a standardized nomenclature system.

First proposed in 1987 and updated on a biannual basis, the nomenclature system is based on divergent evolution of amino acid sequence comparisons between P450 proteins. The P450 genes are divided into families and subfamilies: enzymes within a family display greater than 40% amino acid similarity, and those within the same subfamily display 55% similarity. P450 genes are named with the root symbol CYP followed by an arabic numeral designating the P450 family, a letter denoting the subfamily, and a further arabic numeral designating the individual gene. Thus, CYP1A1 represents P450 gene 1 in family 1 and subfamily A.

As of February 1995, there are 403 CYP genes in the database, composed of 59 families and 105 subfamilies. These include eight lower eukaryotic families, 15 plant families, and 19 bacterial families. The 15 human P450 gene families comprise 26 subfamilies, 22 of which have been mapped to chromosomal locations throughout most of the genome. Some sequences are clearly orthologous across many species — for example, only one CYP17 (steroid 17α-hydroxylase) gene has been found in all vertebrates examined to date; other sequences within a subfamily are highly duplicated, making the identification of orthologous pairs impossible (*e.g.*, the CYP2C subfamily). Interestingly, human and yeast share an orthologous gene in the CYP51 family. Numerous comprehensive reviews are available for readers seeking further information on the P450 superfamily.

The success of the P450 nomenclature system has resulted in similar terminology systems being developed for the UDP glucuronosyltransferases and flavin-containing mono-oxygenases. Similar nomenclature systems based on divergent evolution are also under development for several other drug-metabolizing enzyme superfamilies (*e.g.*, sulphotransferases, epoxide hydrolases and aldehyde dehydrogenases).

Recently, we divided the mammalian P450 gene superfamily into three groups — those involved principally with foreign chemical metabolism, those involved in the synthesis of various steroid hormones, and those participating in other important endogenous functions. It is the xenobiotic-metabolizing P450 enzymes that assume the most significance for prediction of toxicity.

Xenobiotic-metabolizing P450 Enzymes

P450 enzymes involved in the metabolism of foreign compounds and drugs are almost always found within families CYP1, CYP2, CYP3 and CYP4. These P450 enzymes catalyze a wide variety of metabolic reactions, with a single P450 often capable of metabolizing many different compounds. In addition, multiple P450 enzymes may metabolize a single compound at different sites. Also, a compound may be metabolized at the same, single site by several P450s, although at varying rates.

A most important property of the drug-metabolizing P450 enzymes is that many of these genes are inducible by the very substances which serve as their substrates. On the other hand, other P450 genes are induced by nonsubstrates. This phenomenon of enzyme induction underlies many drug-drug interactions of therapeutic importance.

Although present in many tissues, these particular P450 enzymes are found in relatively high levels in the liver, the primary site of drug metabolism. Some of the xenobiotic-metabolizing P450 enzymes exhibit activity toward certain endogenous substrates (e.g., arachidonic acid). However, it is generally believed that most of these xenobiotic-metabolizing P450 enzymes do not play important physiological roles — although this has not been established experimentally as yet. The selective homozygous disruption, or "knock-out," of individual xenobiotic-metabolizing P450 genes by means of gene targeting methodologies in mice is likely to provide unequivocal information soon with regard to physiological roles of the xenobiotic-metabolizing P450s.

In contrast to P450 families encoding enzymes involved primarily in physiological processes, families encoding xenobiotic-metabolizing P450 enzymes display marked species specificity and frequently contain many active genes per subfamily.

Given the apparent lack of physiological substrates, it is possible that P450 enzymes in families CYP1, CYP2, CYP3 and CYP4 that have appeared in the past several hundred million years have evolved as a means of detoxifying foreign chemicals encountered in the environment and diet. Clearly, evolution of the xenobiotic-metabolizing P450s would have occurred over a time period which far precedes the synthesis of most of the synthetic chemicals to which humans are now exposed. The genes in these four gene families may have evolved and diverged in animals due to their exposure to plant metabolites during the last 1.2 billion years — a process descriptively termed "animal-plant warfare". Animal-plant warfare is the phenomenon in which plants developed new chemicals (phytoalexins) as a defence mechanism in order to prevent ingestion by animals, and animals, in turn, responded by developing new P450 genes to accommodate the diversifying substrates. Providing further impetus to this proposal are the recently described examples of plant-insect and plant-fungus chemical warfare involving P450 detoxification of toxic substrates.

The following is a brief introduction to several of the human xenobiotic-metabolizing P450 enzyme polymorphisms in which genetic determinants of toxic response are believed to be of high significance. Until recently, P450 polymorphisms were generally suggested by unexpected variance in patient response to administered therapeutic agents. Several P450 polymorphisms are indeed named according to the drug with which the polymorphism was first identified. More recently, research efforts have focused on identification of the precise P450 enzymes involved in the metabolism of chemicals for which variance is observed and the precise characterization of the P450 genes involved. As described earlier, the measurable activity of a P450 enzyme towards a model chemical can be called the phenotype. Allelic differences in a P450 gene for each individual is termed the P450 genotype. As more and more scrutiny is applied to the analysis of P450 genes, the precise molecular basis of previously documented phenotypic variance is becoming clearer.

The CYP1A Subfamily

The CYP1A subfamily comprises two enzymes in humans and all other mammals: these are designated CYP1A1 and CYP1A2 under standard P450 nomenclature. These enzymes are of considerable interest, because they are involved in the metabolic activation of many procarcinogens and are also induced by several compounds of toxicological concern, including dioxin. For example, CYP1A1 metabolically activates many compounds found in cigarette smoke. CYP1A2 metabolically activates many arylamines — associated with urinary bladder cancer — found in the chemical dye industry. CYP1A2 also metabolically activates 4-(methylnitrosamino)-1-(3-pyridyl)-1-butanone (NNK), a tobacco-derived nitrosamine. CYP1A1 and CYP1A2 are also found at higher levels in the lungs of cigarette smokers, due to induction by polycyclic hydrocarbons present in the smoke. The levels of CYP1A1 and CYP1A2 activity are therefore considered to be important determinants of individual response to many potentially toxic chemicals.

Toxicological interest in the CYP1A subfamily was greatly intensified by a 1973 report correlating the level of CYP1A1 inducibility in cigarette smokers with individual susceptibility to lung cancer. The molecular basis of CYP1A1 and CYP1A2 induction has been a major focus of numerous laboratories. The induction process is mediated by a protein termed the Ah receptor to which dioxins and structurally related chemicals bind. The name Ah is derived from the aryl hydrocarbon nature of many CYP1A inducers. Interestingly, differences in the gene encoding the Ah receptor between strains of mice result in marked differences in chemical response and toxicity. A polymorphism in the Ah receptor gene also appears to occur in humans: approximately one-tenth of the population displays high induction of CYP1A1 and may be at greater risk than the other nine-tenths of the population for development of certain chemically induced cancers. The role of the Ah receptor in the control of enzymes in the CYP1A subfamily, and its role as a determinant of human response to chemical exposure, has been the subject of several recent reviews.

Are there other polymorphisms that might control the level of CYP1A proteins in a cell? A polymorphism in the CYP1A1 gene has also been identified, and this appears to influence lung cancer risk amongst Japanese cigarette smokers, although this same polymorphism does not appear to influence risk in other ethnic groups.

CYP2C19

Variations in the rate at which individuals metabolize the anticonvulsant drug (S)-mephenytoin have been well documented for many years. Between 2% and 5% of Caucasians and as many as 25% of Asians are deficient in this activity and may be at greater risk of toxicity from the drug. This enzyme defect has long been known to involve a member of the human CYP2C subfamily, but the precise molecular basis of this deficiency has been the subject of considerable controversy. The major reason for this difficulty was the six or more genes in the human CYP2C subfamily. It was recently demonstrated, however, that a single-base mutation in the CYP2C19 gene is the primary cause of this deficiency. A simple DNA test, based on the polymerase chain reaction (PCR), has also been developed to identify this mutation rapidly in human populations.

CYP2D6

Perhaps the most extensively characterized variation in a P450 gene is that involving the CYP2D6 gene. More than a dozen examples of mutations, rearrangements and deletions affecting this gene have been described. This polymorphism was first suggested 20 years ago by clinical variability in patients' response to the antihypertensive agent debrisoquine. Alterations in the CYP2D6 gene giving rise to altered enzyme activity are therefore collectively termed the debrisoquine polymorphism.

Prior to the advent of DNA-based studies, individuals had been classified as poor or extensive metabolizers (PMs, EMs) of debrisoquine based on metabolite concentrations in urine samples. It is now clear that alterations in the CYP2D6 gene may result in individuals displaying not only poor or extensive debrisoquine metabolism, but also ultrarapid metabolism. Most alterations in the CYP2D6 gene are associated with partial or total deficiency of enzyme function; however, individuals in two families have recently been described who possess multiple functional copies of the CYP2D6 gene, giving rise to ultrarapid metabolism of CYP2D6 substrates. This remarkable observation provides new insights into the wide spectrum of CYP2D6 activity previously observed in population studies. Alterations in CYP2D6 function are of particular significance, given the more than 30 commonly prescribed drugs metabolized by this enzyme. An individual's CYP2D6 function is therefore a major determinant of both therapeutic and toxic response to administered therapy. Indeed, it has recently been argued that consideration of a patient's CYP2D6 status is necessary for the safe use of both psychiatric and cardiovascular drugs.

The role of the CYP2D6 polymorphism as a determinant of individual susceptibility to human diseases such as lung cancer and Parkinson's disease has also been the subject of intense study. While conclusions are difficult to define given the diverse nature of the study protocols utilized, the majority of studies appear to indicate an association between extensive metabolizers of debrisoquine (EM phenotype) and lung cancer. The reasons for such an association are presently unclear. However, the CYP2D6 enzyme has been shown to metabolize NNK, a tobacco-derived nitrosamine.

As DNA-based assays improve—enabling even more accurate assessment of CYP2D6 status—it is anticipated that the precise relationship of CYP2D6 to disease risk will be clarified. Whereas the extensive metabolizer may be linked with susceptibility to lung cancer, the poor metabolizer (PM phenotype) appears to be associated with Parkinson's disease of unknown cause. Whereas these studies are also difficult to compare, it appears that PM individuals having a diminished capacity to metabolize CYP2D6 substrates (*e.g.*, debrisoquine) have a 2- to 2.5-fold increase in risk of developing Parkinson's disease.

CYP2E1

The CYP2E1 gene encodes an enzyme that metabolizes many chemicals, including drugs and many low-molecular-weight carcinogens. This enzyme is also of interest because it is highly inducible by alcohol and may play a role in liver injury induced by chemicals such as chloroform, vinyl chloride and carbon tetrachloride. The enzyme is primarily found in the liver, and the level of enzyme varies markedly between individuals. Close scrutiny of the CYP2E1 gene has resulted in the identification of several polymorphisms (Nebert and McKinnon 1994). A relationship has been reported between the presence of certain structural variations in the CYP2E1 gene and apparent lowered lung cancer risk in some studies; however, there are clear interethnic differences which require clarification of this possible relationship.

The CYP3A Subfamily

In humans, four enzymes have been identified as members of the CYP3A subfamily due to their similarity in amino acid sequence. The CYP3A enzymes metabolize many commonly prescribed drugs such as erythromycin and cyclosporin. The carcinogenic food contaminant aflatoxin B_1 is also a CYP3A substrate. One member of the human CYP3A subfamily, designated CYP3A4, is the principal P450 in human liver as well as being present in the gastrointestinal tract. As is true for many other P450 enzymes, the level of CYP3A4 is highly variable between individuals. A second enzyme, designated CYP3A5, is found in only approximately 25% of livers; the genetic basis of this finding has not been elucidated. The importance of CYP3A4 or CYP3A5 variability as a factor in genetic determinants of toxic response has not yet been established.

Non-P450 Polymorphisms

Numerous polymorphisms also exist within other xenobiotic-metabolizing enzyme superfamilies (*e.g.*, glutathione transferases, UDP glucuronosyltransferases, para-oxonases, dehydrogenases, N-acetyltransferases and flavin-containing monooxygenases). Because the ultimate toxicity of any P450-generated intermediate is dependent on the efficiency of subsequent Phase II detoxification reactions, the combined role of multiple enzyme polymorphisms is important in determining susceptibility to chemically induced diseases. The metabolic balance between Phase I and Phase II reactions is therefore likely to be a major factor in chemically induced human diseases and genetic determinants of toxic response.

The GSTM1 Gene Polymorphism

A well studied example of a polymorphism in a Phase II enzyme is that involving a member of the glutathione S-transferase enzyme superfamily, designated GST mu or GSTM1. This particular enzyme is of considerable toxicological interest because it appears to be involved in the subsequent detoxification of toxic metabolites produced from chemicals in cigarette smoke by the CYP1A1 enzyme. The identified polymorphism in this glutathione transferase gene involves a total absence of functional enzyme in as many as half of all Caucasians studied. This lack of a Phase II enzyme appears to be associated with increased susceptibility to lung cancer. By grouping individuals on the basis of both variant CYP1A1 genes and the deletion or presence of a functional GSTM1 gene, it has been demonstrated that the risk of developing smoking-induced lung cancer varies significantly (Kawajiri, Watanabe and Hayashi 1994). In particular, individuals displaying one rare CYP1A1 gene alteration, in combination with an absence of the GSTM1 gene, were at higher risk (as much as ninefold) of developing lung cancer when exposed to a relatively low level of cigarette smoke. Interestingly, there appear to be interethnic differences in the significance of variant genes which necessitate further study in order to elucidate the precise role of such alterations in susceptibility to disease (Kalow 1962; Nebert and McKinnon 1994; Kawajiri, Watanabe and Hayashi 1994).

Synergistic Effect of Two or More Polymorphisms on the Toxic Response

A toxic response to an environmental agent may be greatly exaggerated by the combination of two pharmacogenetic defects in the same individual, for example, the combined effects of the N-acetyltransferase (NAT2) polymorphism and the glucose-6-phosphate dehydrogenase (G6PD) polymorphism.

Occupational exposure to arylamines constitutes a grave risk of urinary bladder cancer. Since the elegant studies of Cartwright in 1954, it has become clear that the N-acetylator status is a determinant of azo-dye-induced bladder cancer. There is a highly significant correlation between the slow-acetylator phenotype and the occurrence of bladder cancer, as well as the degree of invasiveness of this cancer in the bladder wall. On the contrary, there is a significant association between the rapid-acetylator phenotype and the incidence of colorectal carcinoma. The N-acetyltransferase (NAT1, NAT2) genes have been cloned and sequenced, and DNA-based assays are now able to detect the more than a dozen allelic variants which account for the slow-acetylator phenotype. The NAT2 gene is polymorphic and responsible for most of the variability in toxic response to environmental chemicals.

Glucose-6-phosphate dehydrogenase (G6PD) is an enzyme critical in the generation and maintenance of NADPH. Low or absent G6PD activity can lead to severe drug- or xenobiotic-induced haemolysis, due to the absence of normal levels of reduced glutathione (GSH) in the red blood cell. G6PD deficiency affects at least 300 million people worldwide. More than 10% of African-American males exhibit the less severe phenotype, while certain Sardinian communities exhibit the more severe "Mediterranean type" at frequencies as high as one in every three persons. The G6PD gene has been cloned and localized to the X chromosome, and numerous diverse point mutations account for the large degree of phenotypic heterogeneity seen in G6PD-deficient individuals.

Thiozalsulphone, an arylamine sulpha drug, was found to cause a bimodal distribution of haemolytic anaemia in the treated population. When treated with certain drugs, individuals with the combination of G6PD deficiency plus the slow-acetylator phenotype are more affected than those with the G6PD deficiency alone or the slow-acetylator phenotype alone. G6PD-deficient slow acetylators are at least 40 times more susceptible than normal-G6PD rapid acetylators to thiozalsulphone-induced haemolysis.

Effect of Genetic Polymorphisms on Exposure Assessment

Exposure assessment and biomonitoring also requires information on the genetic make-up of each individual. Given identical exposure to a hazardous chemical, the level of haemoglobin adducts (or other biomarkers) might vary by two or three orders of magnitude among individuals, depending upon each person's metabolic fingerprint.

The same combined pharmacogenetics has been studied in chemical factory workers in Germany. Haemoglobin adducts among workers exposed to aniline and acetanilide are by far the highest in G6PD-deficient slow acetylators, as compared with the other possible combined pharmacogenetic phenotypes. This study has important implications for exposure assessment. These data demonstrate that, although two individuals might be exposed to the same ambient level of hazardous chemical in the work place, the amount of exposure (*via* biomarkers such as haemoglobin adducts) might be estimated to be two or more orders of magnitude less, due to the underlying genetic predisposition of the individual. Likewise, the resulting risk of an adverse health effect may vary by two or more orders of magnitude.

Table. Haemoglobin adducts in workers exposed to aniline and acetanilide

Acetylator status		G6PD deficiency		
Fast	Slow	No	Yes	Hgb adducts
+		+		2
+			+	30
	+	+		20
	+		+	100

Genetic Differences In Binding as well as Metabolism

It should be emphasized that the same case made here for metabolism can also be made for binding. Heritable differences in the binding of environmental agents will greatly affect the toxic response. For example, differences in the mouse cdm gene can profoundly affect individual sensitivity to cadmium induced testicular necrosis Differences in the binding affinity of the Ah receptor are likely affect dioxin-induced toxicity and cancer.Toxic agents, as they exist in the environment or following metabolism or binding, elicit their effects by either a genotoxic pathway (in which damage to DNA occurs) or a non-genotoxic pathway (in which DNA damage and mutagenesis need not occur). Interestingly, it has recently become clear that "classical" DNA-damaging agents can operate *via* a reduced glutathione (GSH)-dependent nongenotoxic signal transduction pathway, which is initiated on or near the cell surface in the absence of DNA and outside the cell nucleus. Genetic differences in metabolism and binding remain, however, as the major determinants in controlling different individual toxic responses.

The General Means by which Toxicity Occurs

Role of Drug-metabolizing Enzymes in Cellular Function

Genetically based variation in drug-metabolizing enzyme function is of major importance in determining individual response to chemicals. These enzymes are pivotal in determining the fate and time course of a foreign chemical following exposure.

The importance of drug-metabolizing enzymes in individual susceptibility to chemical exposure may in fact present a far more complex issue than is evident from this simple discussion of xenobiotic metabolism. In other words, during the past two decades, genotoxic mechanisms (measurements of DNA adducts and protein adducts) have been greatly emphasized. However, what if nongenotoxic mechanisms are at least as important as genotoxic mechanisms in causing toxic responses?

As mentioned earlier, the physiological roles of many drug-metabolizing enzymes involved in xenobiotic metabolism have not been accurately defined. Nebert (1994) has proposed that, because of their presence on this planet for more than 3.5 billion years, drug-metabolizing enzymes were originally (and are now still primarily) responsible for regulating the cellular levels of many nonpeptide ligands important in the transcriptional activation of genes affecting growth, differentiation, apoptosis, homeostasis and neuroendocrine functions. Furthermore, the toxicity of most, if not all, environmental agents occurs by means of agonist or antagonist action on these signal transduction pathways (Nebert 1994). Based on this hypothesis, genetic variability in drug-metabolizing enzymes may have quite dramatic effects on many critical biochemical processes within the cell, thereby leading to important differences in toxic response. It is indeed possible that such a scenario may also underlie many idiosyncratic adverse reactions encountered in patients using commonly prescribed drugs.

CONCLUSIONS

The past decade has seen remarkable progress in our understanding of the genetic basis of differential response to chemicals in drugs, foods and environmental pollutants. Drug-metabolizing enzymes have a profound influence on the way humans respond to chemicals. As our awareness of drug-metabolizing enzyme multiplicity continues to evolve, we are increasingly able to make improved assessments of toxic risk for many drugs and environmental chemicals. This is perhaps most clearly illustrated in the case of the CYP2D6 cytochrome P450 enzyme. Using relatively simple DNA-based tests, it is possible to predict the likely response of any drug predominantly metabolized by this enzyme; this prediction will ensure the safer use of valuable, yet potentially toxic, medication.

The future will no doubt see an explosion in the identification of further polymorphisms (phenotypes) involving drug-metabolizing enzymes. This information will be accompanied by improved, minimally invasive DNA-based tests to identify genotypes in human populations.

Such studies should be particularly informative in evaluating the role of chemicals in the many environmental diseases of presently unknown origin. The consideration of multiple drug-metabolizing enzyme polymorphisms, in combination, is also likely to represent a particularly fertile research area. Such studies will clarify the role of chemicals in the causation of cancers. Collectively, this information should enable the formulation of increasingly individualized advice on avoidance of chemicals likely to be of individual concern. This is the field of preventive toxicology. Such advice will no doubt greatly assist all individuals in coping with the ever increasing chemical burden to which we are exposed.

INTRODUCTION AND CONCEPTS

Mechanistic toxicology is the study of how chemical or physical agents interact with living organisms to cause toxicity. Knowledge of the mechanism of toxicity of a substance enhances the ability to prevent toxicity and design more desirable chemicals; it constitutes the basis for therapy upon overexposure, and frequently enables a further understanding of fundamental biological processes. For purposes of this Encyclopaedia the emphasis will be placed on animals to predict human toxicity. Different areas of toxicology include mechanistic, descriptive, regulatory, forensic and environmental toxicology. All of these benefit from understanding the fundamental mechanisms of toxicity.

Why Understand Mechanisms of Toxicity?

Understanding the mechanism by which a substance causes toxicity enhances different areas of toxicology in different ways. Mechanistic understanding helps the governmental regulator to establish legally binding safe limits for human exposure. It helps toxicologists in recommending courses of action regarding clean-up or remediation of contaminated sites and, along with physical and chemical properties of the substance or mixture, can be used to select the degree of protective equipment required. Mechanistic knowledge is also useful in forming the basis for therapy and the design of new drugs for treatment of human disease. For the forensic toxicologist the mechanism of toxicity often provides insight as to how a chemical or physical agent can cause death or incapacitation.

If the mechanism of toxicity is understood, descriptive toxicology becomes useful in predicting the toxic effects of related chemicals. It is important to understand, however, that a lack of mechanistic information does not deter health professionals from protecting human health. Prudent decisions based on animal studies and human experience are used to establish safe exposure levels. Traditionally, a margin of safety was established by using the "no adverse effect level" or a "lowest adverse effect level" from animal studies (using repeated-exposure designs) and dividing that level by a factor of 100 for occupational exposure or 1,000 for other human environmental exposure. The success of this process is evident from the few incidents of adverse health effects attributed to chemical exposure in workers where appropriate exposure limits had been set and adhered to in the past. In addition, the human lifespan continues to increase, as does the quality of life. Overall the use of toxicity data has led to effective regulatory and voluntary control. Detailed knowledge of toxic mechanisms will enhance the predictability of newer risk models currently being developed and will result in continuous improvement.

Understanding environmental mechanisms is complex and presumes a knowledge of ecosystem disruption and homeostasis (balance). While not discussed in this article, an enhanced understanding of toxic mechanisms and their ultimate consequences in an ecosystem would help scientists to make prudent decisions regarding the handling of municipal and industrial waste material. Waste management is a growing area of research and will continue to be very important in the future.

Techniques for Studying Mechanisms of Toxicity

The majority of mechanistic studies start with a descriptive toxicological study in animals or clinical observations in humans. Ideally, animal studies include careful behavioural and clinical observations, careful biochemical examination of elements of the blood and urine for signs of adverse function of major biological systems in the body, and a post-mortem evaluation of all organ systems by microscopic examination to check for injury (see OECD test guidelines; EC directives on chemical evaluation; US EPA test rules; Japan chemicals regulations). This is analogous to a thorough human physical examination that would take place in a hospital over a two- to three-day time period except for the post-mortem examination.

Understanding mechanisms of toxicity is the art and science of observation, creativity in the selection of techniques to test various hypotheses, and innovative integration of signs and symptoms into a causal relationship. Mechanistic studies start with exposure, follow the time-related distribution and fate in the body (pharmacokinetics), and measure the resulting toxic effect at some level of the system and at some dose level. Different substances can act at different levels of the biological system in causing toxicity.

Exposure

The route of exposure in mechanistic studies is usually the same as for human exposure. Route is important because there can be effects that occur locally at the site of exposure in addition to systemic effects after the chemical has been absorbed into the blood and distributed throughout the body. A simple yet cogent example of a local effect would be irritation and eventual corrosion of the skin following application of strong acid or alkaline solutions designed for cleaning hard surfaces. Similarly, irritation and cellular death can occur in cells lining the nose and/or lungs following exposure to irritant vapours or gases such as oxides of nitrogen or ozone. (Both are constituents of air pollution, or smog). Following absorption of a chemical into blood through the skin, lungs or gastrointestinal tract, the concentration in any organ or tissue is controlled by many factors which determine the pharmacokinetics of the chemical in the body. The body has the ability to activate as well as detoxify various chemicals as noted below.

Role of Pharmacokinetics in Toxicity

Pharmacokinetics describes the time relationships for chemical absorption, distribution, metabolism (biochemical alterations in the body) and elimination or excretion from the body. Relative to mechanisms of toxicity, these pharmacokinetic variables can be very important and in some instances determine whether toxicity will or will not occur. For instance, if a material is not absorbed in a sufficient amount, systemic toxicity (inside the body) will not occur. Conversely, a highly reactive chemical that is detoxified quickly (seconds or minutes) by digestive or liver enzymes may not have the time to cause toxicity. Some polycyclic halogenated substances and mixtures as well as certain metals like lead would not cause significant toxicity if excretion were rapid; but accumulation to sufficiently high levels determines their toxicity since excretion is not rapid (sometimes measured in years). Fortunately, most chemicals do not have such long retention in the body. Accumulation of an innocuous material still would not induce toxicity. The rate of elimination from the body and detoxication is frequently referred to as the half-life of the chemical, which is the time for 50% of the chemical to be excreted or altered to a non-toxic form.

However, if a chemical accumulates in a particular cell or organ, that may signal a reason to further examine its potential toxicity in that organ. More recently, mathematical models have been developed to extrapolate pharmacokinetic variables from animals to humans. These pharmacokinetic models are extremely useful in generating hypotheses and testing whether the experimental animal may be a good representation for humans. Numerous chapters and texts have been written on this subject.

Different Levels and Systems Can Be Adversely Affected

Toxicity can be described at different biological levels. Injury can be evaluated in the whole person (or animal), the organ system, the cell or the molecule.

Organ systems include the immune, respiratory, cardiovascular, renal, endocrine, digestive, muscolo-skeletal, blood, reproductive and central nervous systems. Some key organs include the liver, kidney, lung, brain, skin, eyes, heart, testes or ovaries, and other major organs. At the cellular/biochemical level, adverse effects include interference with normal protein function, endocrine receptor function, metabolic energy inhibition, or xenobiotic (foreign substance) enzyme inhibition or induction. Adverse effects at the molecular level include alteration of the normal function of DNA-RNA transcription, of specific cytoplasmic and nuclear receptor binding, and of genes or gene products.

Ultimately, dysfunction in a major organ system is likely caused by a molecular alteration in a particular target cell within that organ. However, it is not always possible to trace a mechanism back to a molecular origin of causation, nor is it necessary. Intervention and therapy can be designed without a complete understanding of the molecular target. However, knowledge about the specific mechanism of toxicity increases the predictive value and accuracy of extrapolation to other chemicals. The arrows indicate that the consequences to an individual can be determined from top down (exposure, pharmacokinetics to system/organ toxicity) or from bottom up (molecular change, cellular/biochemical effect to system/organ toxicity).

Examples of Mechanisms of Toxicity

Mechanisms of toxicity can be straightforward or very complex. Frequently, there is a difference among the type of toxicity, the mechanism of toxicity, and the level of effect, related to whether the adverse effects are due to a single, acute high dose (like an accidental poisoning), or a lower-dose repeated exposure (from occupational or environmental exposure). Classically, for testing purposes, an acute, single high dose is given by direct intubation into the stomach of a rodent or exposure to an atmosphere of a gas or vapour for two to four hours, whichever best resembles the human exposure. The animals are observed over a two-week period following exposure and then the major external and internal organs are examined for injury. Repeated-dose testing ranges from months to years. For rodent species, two years is considered a chronic (lifetime) study sufficient to evaluate toxicity and carcinogenicity, whereas for non-human primates, two years would be considered a subchronic (less than lifetime) study to evaluate repeated dose toxicity. Following exposure a complete examination of all tissues, organs and fluids is conducted to determine any adverse effects.

Acute Toxicity Mechanisms

The following examples are specific to high-dose, acute effects which can lead to death or severe incapacitation. However, in some cases, intervention will result in transient and fully reversible effects. The dose or severity of exposure will determine the result.

Simple asphyxiants. The mechanism of toxicity for inert gases and some other non-reactive substances is lack of oxygen (anoxia). These chemicals, which cause deprivation of oxygen to the central nervous system (CNS), are termed simple asphyxiants. If a person enters a closed space that contains nitrogen without sufficient oxygen, immediate oxygen depletion occurs in the brain and leads to unconsciousness and eventual death if the person is not rapidly removed. In extreme cases (near zero oxygen) unconsciousness can occur in a few seconds. Rescue depends on rapid removal to an oxygenated environment. Survival with irreversible brain damage can occur from delayed rescue, due to the death of neurons, which cannot regenerate.

Chemical asphyxiants. Carbon monoxide (CO) competes with oxygen for binding to haemoglobin (in red blood cells) and therefore deprives tissues of oxygen for energy metabolism; cellular death can result. Intervention includes removal from the source of CO and treatment with oxygen. The direct use of oxygen is based on the toxic action of CO. Another potent chemical asphyxiant is cyanide. The cyanide ion interferes with cellular metabolism and utilization of oxygen for energy. Treatment with sodium nitrite causes a change in haemoglobin in red blood cells to methaemoglobin. Methaemoglobin has a greater binding affinity to the cyanide ion than does the cellular target of cyanide. Consequently, the methaemoglobin binds the cyanide and keeps the cyanide away from the target cells. This forms the basis for antidotal therapy.

Central nervous system (CNS) depressants. Acute toxicity is characterized by sedation or unconsciousness for a number of materials like solvents which are not reactive or which are transformed to reactive intermediates. It is hypothesized that sedation/anaesthesia is due to an interaction of the solvent with the membranes of cells in the CNS, which impairs their ability to transmit electrical and chemical signals. While sedation may seem a mild form of toxicity and was the basis for development of the early anaesthetics, "the dose still makes the poison". If sufficient dose is administered by ingestion or inhalation the animal can die due to respiratory arrest. If anaesthetic death does not occur, this type of toxicity is usually readily reversible when the subject is removed from the environment or the chemical is redistributed or eliminated from the body.

Skin effects. Adverse effects to the skin can range from irritation to corrosion, depending on the substance encountered. Strong acids and alkaline solutions are incompatible with living tissue and are corrosive, causing chemical burns and possible scarring. Scarring is due to death of the dermal, deep skin cells responsible for regeneration. Lower concentrations may just cause irritation of the first layer of skin.

Another specific toxic mechanism of skin is that of chemical sensitization. As an example, sensitization occurs when 2,4-dinitrochlorobenzene binds with natural proteins in the skin and the immune system recognizes the altered protein-bound complex as a foreign material. In responding to this foreign material, the immune system activates special cells to eliminate the foreign substance by release of mediators (cytokines) which cause a rash or dermatitis (see "Immunotoxicology"). This is the same reaction of the immune system when exposure to poison ivy occurs. Immune sensitization is very specific to the particular chemical and takes at least two exposures before a response is elicited. The first exposure sensitizes (sets up the cells to recognize the chemical), and subsequent exposures trigger the immune system response. Removal from contact and symptomatic therapy with steroid-containing anti-inflammatory creams are usually effective in treating sensitized individuals. In serious or refractory cases a systemic acting immunosuppresant like prednisone is used in conjunction with topical treatment.

Lung sensitization. An immune sensitization response is elicited by toluene diisocyanate (TDI), but the target site is the lungs. TDI over-exposure in susceptible individuals causes lung oedema (fluid build-up), bronchial constriction and impaired breathing. This is a serious condition and requires removing the individual from potential subsequent exposures. Treatment is primarily symptomatic. Skin and lung sensitization follow a dose response. Exceeding the level set for occupational exposure can cause adverse effects.

Eye effects. Injury to the eye ranges from reddening of the outer layer (swimming-pool redness) to cataract formation of the cornea to damage to the iris (coloured part of the eye). Eye irritation tests are conducted when it is believed serious injury will not occur. Many of the mechanisms causing skin corrosion can also cause injury to the eyes. Materials corrosive to the skin, like strong acids (pH less than 2) and alkali (pH greater than 11.5), are not tested in the eyes of animals because most will cause corrosion and blindness due to a mechanism similar to that which causes skin corrosion. In addition, surface active agents like detergents and surfactants can cause eye injury ranging from irritation to corrosion. A group of materials that requires caution is the positively charged (cationic) surfactants, which can cause burns, permanent opacity of the cornea and vascularization (formation of blood vessels). Another chemical, dinitrophenol, has a specific effect of cataract formation. This appears to be related to concentration of this chemical in the eye, which is an example of pharmacokinetic distributional specificity.

While the listing above is far from exhaustive, it is designed to give the reader an appreciation for various acute toxicity mechanisms.

Subchronic and Chronic Toxicity Mechanisms

When given as a single high dose, some chemicals do not have the same mechanism of toxicity as when given repeatedly as a lower but still toxic dose. When a single high dose is given, there is always the possibility of exceeding the person's ability to detoxify or excrete the chemical, and this can lead to a different toxic response than when lower repetitive doses are given. Alcohol is a good example. High doses of alcohol lead to primary central nervous system effects, while lower repetitive doses result in liver injury.

Anticholinesterase inhibition. Most organophosphate pesticides, for example, have little mammalian toxicity until they are metabolically activated, primarily in the liver. The primary mechanism of action of organophosphates is the inhibition of acetylcholinesterase (AChE) in the brain and peripheral nervous system. AChE is the normal enzyme that terminates the stimulation of the neurotransmitter acetylcholine. Slight inhibition of AChE over an extended period has not been associated with adverse effects. At high levels of exposure, inability to terminate this neuronal stimulation results in overstimulation of the cholinergic nervous system. Cholinergic overstimulation ultimately results in a host of symptoms, including respiratory arrest, followed by death if not treated. The primary treatment is the administration of atropine, which blocks the effects of acetylcholine, and the administration of pralidoxime chloride, which reactivates the inhibited AChE. Therefore, both the cause and the treatment of organophosphate toxicity are addressed by understanding the biochemical basis of toxicity.

Metabolic activation. Many chemicals, including carbon tetrachloride, chloroform, acetylaminofluorene, nitrosamines, and paraquat are metabolically activated to free radicals or other reactive intermediates which inhibit and interfere with normal cellular function. At high levels of exposure this results in cell death (see "Cellular injury and cellular death"). While the specific interactions and cellular targets remain unknown, the organ systems which have the capability to activate these chemicals, like the liver, kidney and lung, are all potential targets for injury. Specifically, particular cells within an organ have a greater or lesser capacity to activate or detoxify these intermediates, and this capacity determines the intracellular susceptibility within an organ. Metabolism is one reason why an understanding of pharmacokinetics, which describes these types of transformations and the distribution and elimination of these intermediates, is important in recognizing the mechanism of action of these chemicals.

Cancer mechanisms. Cancer is a multiplicity of diseases, and while the understanding of certain types of cancer is increasing rapidly due to the many molecular biological techniques that have been developed since 1980, there is still much to learn. However, it is clear that cancer development is a multi-stage process, and critical genes are key to different types of cancer. Alterations in DNA (somatic mutations) in a number of these critical genes can cause increased susceptibility or cancerous lesions (see "Genetic toxicology"). Exposure to natural chemicals (in cooked foods like beef and fish) or synthetic chemicals (like benzidine, used as a dye) or physical agents (ultraviolet light from the sun, radon from soil, gamma radiation from medical procedures or industrial activity) are all contributors to somatic gene mutations. However, there are natural and synthetic substances (such as anti-oxidants) and DNA repair processes which are protective and maintain homeostasis. It is clear that genetics is an important factor in cancer, since genetic disease syndromes such as xeroderma pigmentosum, where there is a lack of normal DNA repair, dramatically increase susceptibility to skin cancer from exposure to ultraviolet light from the sun.

Reproductive mechanisms. Similar to cancer, many mechanisms of reproductive and/or developmental toxicity are known, but much is to be learned. It is known that certain viruses (such as rubella), bacterial infections and drugs (such as thalidomide and vitamin A) will adversely affect development. Recently, work by Khera (1991), reviewed by Carney (1994), show good evidence that the abnormal developmental effects in animal tests with ethylene glycol are attributable to maternal metabolic acidic metabolites. This occurs when ethylene glycol is metabolized to acid metabolites including glycolic and oxalic acid. The subsequent effects on the placenta and foetus appear to be due to this metabolic toxication process.

Conclusion

The intent of this article is to give a perspective on several known mechanisms of toxicity and the need for future study. It is important to understand that mechanistic knowledge is not absolutely necessary to protect human or environmental health. This knowledge will enhance the professional's ability to better predict and manage toxicity. The actual techniques used in elucidating any particular mechanism depend upon the collective knowledge of the scientists and the thinking of those who make decisions regarding human health.

CELLULAR INJURY AND CELLULAR DEATH

Virtually all of medicine is devoted to either preventing cell death, in diseases such as myocardial infarction, stroke, trauma and shock, or causing it, as in the case of infectious diseases and cancer. It is, therefore, essential to understand the nature and mechanisms involved. Cell death has been classified as "accidental", that is, caused by toxic agents, ischaemia and so on, or "programmed", as occurs during embryological development, including formation of digits, and resorption of the tadpole tail.

Cell injury and cell death are, therefore, important both in physiology and in pathophysiology. Physiological cell death is extremely important during embryogenesis and embryonic development. The study of cell death during development has led to important and new information on the molecular genetics involved, especially through the study of development in invertebrate animals. In these animals, the precise location and the significance of cells that are destined to undergo cell death have been carefully studied and, with the use of classic mutagenesis techniques, several involved genes have now been identified. In adult organs, the balance between cell death and cell proliferation controls organ size. In some organs, such as the skin and the intestine, there is a continual turnover of cells. In the skin, for example, cells differentiate as they reach the surface, and finally undergo terminal differentiation and cell death as keratinization proceeds with the formation of crosslinked envelopes.

Many classes of toxic chemicals are capable of inducing acute cell injury followed by death. These include anoxia and ischaemia and their chemical analogues such as potassium cyanide; chemical carcinogens, which form electrophiles that covalently bind to proteins in nucleic acids; oxidant chemicals, resulting in free radical formation and oxidant injury; activation of complement; and a variety of calcium ionophores. Cell death is also an important component of chemical carcinogenesis; many complete chemical carcinogens, at carcinogenic doses, produce acute necrosis and inflammation followed by regeneration and preneoplasia.

DEFINITIONS

Cell Injury

Cell injury is defined as an event or stimulus, such as a toxic chemical, that perturbs the normal homeostasis of the cell, thus causing a number of events to occur. The principal targets of lethal injury illustrated are inhibition of ATP synthesis, disruption of plasma membrane integrity or withdrawal of essential growth factors.

Lethal injuries result in the death of a cell after a variable period of time, depending on temperature, cell type and the stimulus; or they can be sublethal or chronic—that is, the injury results in an altered homeostatic state which, though abnormal, does not result in cell death. In the case of a lethal injury, there is a phase prior to the time of cell death called the "prelethal phase". If the injurious stimulus, such as anoxia, can be removed during this time, the cell will recover; however, after a particular point in time (the "point of no return" or point of cell death), the removal of the injury does not result in recovery but instead the cell undergoes degradation and hydrolysis, ultimately reaching physical-chemical equilibrium with the environment. This is the phase known as necrosis. During the prelethal phase, several principal types of change occur, depending on the cell and the type of injury. These are known as apoptosis and oncosis.

Apoptosis

Apoptosis is derived from the Greek words apo, meaning away from, and ptosis, meaning to fall. The term falling away from is derived from the fact that, during this type of prelethal change, the cells shrink and undergo marked blebbing at the periphery. The blebs then detach and float away. Apoptosis occurs in a variety of cell types following various types of toxic injury (Wyllie, Kerr and Currie 1980). It is especially prominent in lymphocytes, where it is the predominant mechanism for turnover of lymphocyte clones. The resulting fragments result in the basophilic bodies seen within macrophages in lymph nodes. In other organs, apoptosis typically occurs in single cells which are rapidly cleared away before and following death by phagocytosis of the fragments by adjacent parenchymal cells or by macrophages. Apoptosis occurring in single cells with subsequent phagocytosis typically does not result in inflammation. Prior to death, apoptotic cells show a very dense cytosol with normal or condensed mitochondria. The endoplasmic reticulum (ER) is normal or only slightly dilated. The nuclear chromatin is markedly clumped along the nuclear envelope and around the nucleolus. The nuclear contour is also irregular and nuclear fragmentation occurs. The chromatin condensation is associated with DNA fragmentation which, in many instances, occurs between nucleosomes, giving a characteristic ladder appearance on electrophoresis.

In apoptosis, increased $[Ca^{2+}]_i$ may stimulate K^+ efflux resulting in cell shrinkage, which probably requires ATP. Injuries that totally inhibit ATP synthesis, therefore, are more likely to result in apoptosis. A sustained increase of $[Ca^{2+}]_i$ has a number of deleterious effects including activation of proteases, endonucleases, and phospholipases. Endonuclease activation results in single and double DNA strand breaks which, in turn, stimulate increased levels of p53 and in poly-ADP ribosylation, and of nuclear proteins which are essential in DNA repair. Activation of proteases modifies a number of substrates including actin and related proteins leading to bleb formation. Another important substrate is poly(ADP-ribose) polymerase (PARP), which inhibits DNA repair. Increased $[Ca^{2+}]_i$ is also associated with activation of a number of protein kinases, such as MAP kinase, calmodulin kinase and others. Such kinases are involved in activation of transcription factors which initiate transcription of immediate-early genes, for example, c-fos, c-jun and c-myc, and in activation of phospholipase A_2 which results in permeabilization of the plasma membrane and of intracellular membranes such as the inner membrane of mitochondria.

Oncosis

Oncosis, derived from the Greek word onkos, to swell, is so named because in this type of prelethal change the cell begins to swell almost immediately following the injury (Majno and Joris 1995). The reason for the swelling is an increase in cations in the water within the cell. The principal cation responsible is sodium, which is normally regulated to maintain cell volume. However, in the absence of ATP or if Na-ATPase of the plasmalemma is inhibited, volume control is lost because of intracellular protein, and sodium in the water continuing to increase. Among the early events in oncosis are, therefore, increased $[Na^+]_i$ which leads to cellular swelling and increased $[Ca^{2+}]_i$ resulting either from influx from the extracellular space or release from intracellular stores. This results in swelling of the cytosol, swelling of the endoplasmic reticulum and Golgi apparatus, and the formation of watery blebs around the cell surface. The mitochondria initially undergo condensation, but later they too show high-amplitude swelling because of damage to the inner mitochondrial membrane. In this type of prelethal change, the chromatin undergoes condensation and ultimately degradation; however, the characteristic ladder pattern of apoptosis is not seen.

Necrosis

Necrosis refers to the series of changes that occur following cell death when the cell is converted to debris which is typically removed by the inflammatory response. Two types can be distinguished: oncotic necrosis and apoptotic necrosis. Oncotic necrosis typically occurs in large zones, for example, in a myocardial infarct or regionally in an organ after chemical toxicity, such as the renal proximal tubule following administration of $HgCl_2$. Broad zones of an organ are involved and the necrotic cells rapidly incite an inflammatory reaction, first acute and then chronic. In the event that the organism survives, in many organs necrosis is followed by clearing away of the dead cells and regeneration, for example, in the liver or kidney following chemical toxicity. In contrast, apoptotic necrosis typically occurs on a single cell basis and the necrotic debris is formed within the phagocytes of macrophages or adjacent parenchymal cells. The earliest characteristics of necrotic cells include interruptions in plasma membrane continuity and the appearance of flocculent densities, representing denatured proteins within the mitochondrial matrix. In some forms of injury that do not initially interfere with mitochondrial calcium accumulation, calcium phosphate deposits can be seen within the mitochondria. Other membrane systems are similarly fragmenting, such as the ER, the lysosomes and the Golgi apparatus. Ultimately, the nuclear chromatin undergoes lysis, resulting from attack by lysosomal hydrolases. Following cell death, lysosomal hydrolases play an important part in clearing away debris with cathepsins, nucleolases and lipases since these have an acid pH optimum and can survive the low pH of necrotic cells while other cellular enzymes are denatured and inactivated.

MECHANISMS

Initial Stimulus

In the case of lethal injuries, the most common initial interactions resulting in injury leading to cell death are interference with energy metabolism, such as anoxia, ischaemia or inhibitors of respiration, and glycolysis such as potassium cyanide, carbon monoxide, iodo-acetate, and so on. As mentioned above, high doses of compounds that inhibit energy metabolism typically result in oncosis. The other common type of initial injury resulting in acute cell death is modification of the function of the plasma membrane (Trump and Arstila 1971; Trump, Berezesky and Osornio-Vargas 1981). This can either be direct damage and permeabilization, as in the case of trauma or activation of the C5b-C9 complex of complement, mechanical damage to the cell membrane or inhibition of the sodium-potassium (Na^+-K^+) pump with glycosides such as ouabain. Calcium ionophores such as ionomycin or A23187, which rapidly carry [Ca^{2+}] down the gradient into the cell, also cause acute lethal injury. In some cases, the pattern in the prelethal change is apoptosis; in others, it is oncosis.

Signalling Pathways

With many types of injury, mitochondrial respiration and oxidative phosphorylation are rapidly affected. In some cells, this stimulates anaerobic glycolysis, which is capable of maintaining ATP, but with many injuries this is inhibited. The lack of ATP results in failure to energize a number of important homeostatic processes, in particular, control of intracellular ion homeostasis (Trump and Berezesky 1992; Trump, Berezesky and Osornio-Vargas 1981). This results in rapid increases of [Ca^{2+}]$_i$, and increased [Na^+] and [Cl^-] results in cell swelling. Increases in [Ca^{2+}]$_i$ result in the activation of a number of other signalling mechanisms discussed below, including a series of kinases, which can result in increased immediate early gene transcription. Increased $[Ca^{2+}]_i$ also modifies cytoskeletal function, in part resulting in bleb formation and in the activation of endonucleases, proteases and phospholipases. These seem to trigger many of the important effects discussed above, such as membrane damage through protease and lipase activation, direct degradation of DNA from endonuclease activation, and activation of kinases such as MAP kinase and calmodulin kinase, which act as transcription factors.

Through extensive work on development in the invertebrate C. elegans and Drosophila, as well as human and animal cells, a series of pro-death genes have been identified. Some of these invertebrate genes have been found to have mammalian counterparts. For example, the ced-3 gene, which is essential for programmed cell death in C. elegans, has protease activity and a strong homology with the mammalian interleukin converting enzyme (ICE). A closely related gene called apopain or prICE has recently been identified with even closer homology. In Drosophila, the reaper gene seems to be involved in a signal that leads to programmed cell death. Other pro-death genes include the Fas membrane protein and the important tumour-suppressor gene, p53, which is widely conserved. p53 is induced at the protein level following DNA damage and when phosphorylated acts as a transcription factor for other genes such as gadd45 and waf-1, which are involved in cell death signalling. Other immediate early genes such as c-fos, c-jun, and c-myc also seem to be involved in some systems.

At the same time, there are anti-death genes which appear to counteract the pro-death genes. The first of these to be identified was ced-9 from C. elegans, which is homologous to bcl-2 in humans. These genes act in an as yet unknown way to prevent cell killing by either genetic or chemical toxins. Some recent evidence indicates that bcl-2 may act as an antioxidant. Currently, there is much effort underway to develop an understanding of the genes involved and to develop ways to activate or inhibit these genes, depending on the situation.

GENETIC TOXICOLOGY

Genetic toxicology, by definition, is the study of how chemical or physical agents affect the intricate process of heredity. Genotoxic chemicals are defined as compounds that are capable of modifying the hereditary material of living cells. The probability that a particular chemical will cause genetic damage inevitably depends on several variables, including the organism's level of exposure to the chemical, the distribution and retention of the chemical once it enters the body, the efficiency of metabolic activation and/or detoxification systems in target tissues, and the reactivity of the chemical or its metabolites with critical macromolecules within cells. The probability that genetic damage will cause disease ultimately depends on the nature of the damage, the cell's ability to repair or amplify genetic damage, the opportunity for expressing whatever alteration has been induced, and the ability of the body to recognize and suppress the multiplication of aberrant cells.

In higher organisms, hereditary information is organized in chromosomes. Chromosomes consist of tightly condensed strands of protein-associated DNA. Within a single chromosome, each DNA molecule exists as a pair of long, unbranched chains of nucleotide subunits linked together by phosphodiester bonds that join the 5 carbon of one deoxyribose moiety to the 3 carbon of the next. In addition, one of four different nucleotide bases (adenine, cytosine, guanine or thymine) is attached to each deoxyribose subunit like beads on a string. Three-dimensionally, each pair of DNA strands forms a double helix with all of the bases oriented toward the inside of the spiral. Within the helix, each base is associated with its complementary base on the opposite DNA strand; hydrogen bonding dictates strong, noncovalent pairing of adenine with thymine and guanine with cytosine. Since the sequence of nucleotide bases is complementary throughout the entire length of the duplex DNA molecule, both strands carry essentially the same genetic information. In fact, during DNA replication each strand serves as a template for the production of a new partner strand.

Using RNA and an array of different proteins, the cell ultimately deciphers the information encoded by the linear sequence of bases within specific regions of DNA (genes) and produces proteins that are essential for basic cell survival as well as normal growth and differentiation. In essence, the nucleotides function like a biological alphabet which is used to code for amino acids, the building blocks of proteins.

When incorrect nucleotides are inserted or nucleotides are lost, or when unnecessary nucleotides are added during DNA synthesis, the mistake is called a mutation. It has been estimated that less than one mutation occurs for every 10^9 nucleotides incorporated during the normal replication of cells. Although mutations are not necessarily harmful, alterations causing inactivation or overexpression of important genes can result in a variety of disorders, including cancer, hereditary disease, developmental abnormalities, infertility and embryonic or perinatal death. Very rarely, a mutation can lead to enhanced survival; such occurrences are the basis of natural selection.

Although some chemicals react directly with DNA, most require metabolic activation. In the latter case, electrophilic intermediates such as epoxides or carbonium ions are ultimately responsible for inducing lesions at a variety of nucleophilic sites within the genetic material. In other instances, genotoxicity is mediated by by-products of compound interaction with intracellular lipids, proteins, or oxygen.

Because of their relative abundance in cells, proteins are the most frequent target of toxicant interaction. However, modification of DNA is of greater concern due to the central role of this molecule in regulating growth and differentiation through multiple generations of cells.

At the molecular level, electrophilic compounds tend to attack oxygen and nitrogen in DNA. Although oxygens within phosphate groups in the DNA backbone are also targets for chemical modification, damage to bases is thought to be biologically more relevant since these groups are considered to be the primary informational elements in the DNA molecule.

Compounds that contain one electrophilic moiety typically exert genotoxicity by producing mono-adducts in DNA. Similarly, compounds that contain two or more reactive moieties can react with two different nucleophilic centres and thereby produce intra- or inter-molecular crosslinks in genetic material. Interstrand DNA-DNA and DNA-protein crosslinks can be particularly cytotoxic since they can form complete blocks to DNA replication. For obvious reasons, the death of a cell eliminates the possibility that it will be mutated or neoplastically transformed. Genotoxic agents can also act by inducing breaks in the phosphodiester backbone, or between bases and sugars (producing abasic sites) in DNA. Such breaks may be a direct result of chemical reactivity at the damage site, or may occur during the repair of one of the aforementioned types of DNA lesion.

Over the past thirty to forty years, a variety of techniques have been developed to monitor the type of genetic damage induced by various chemicals. Such assays are described in detail elsewhere in this chapter and Encyclopaedia.

Misreplication of "microlesions" such as mono-adducts, abasic sites or single-strand breaks may ultimately result in nucleotide base-pair substitutions, or the insertion or deletion of short polynucleotide fragments in chromosomal DNA. In contrast, "macrolesions," such as bulky adducts, crosslinks, or double-strand breaks may trigger the gain, loss or rearrangement of relatively large pieces of chromosomes. In any case, the consequences can be devastating to the organism since any one of these events can lead to cell death, loss of function or malignant transformation of cells. Exactly how DNA damage causes cancer is largely unknown. It is currently believed the process may involve inappropriate activation of proto-oncogenes such as myc and ras, and/or inactivation of recently identified tumour suppressor genes such as p53. Abnormal expression of either type of gene abrogates normal cellular mechanisms for controlling cell proliferation and/or differentiation.

The preponderance of experimental evidence indicates that the development of cancer following exposure to electrophilic compounds is a relatively rare event. This can be explained, in part, by the cell's intrinsic ability to recognize and repair damaged DNA or the failure of cells with damaged DNA to survive. During repair, the damaged base, nucleotide or short stretch of nucleotides surrounding the damage site is removed and (using the opposite strand as a template) a new piece of DNA is synthesized and spliced into place. To be effective, DNA repair must occur with great accuracy prior to cell division, before opportunities for the propagation of mutation.

Clinical studies have shown that people with inherited defects in the ability to repair damaged DNA frequently develop cancer and/or developmental abnormalities at an early age. Such examples provide strong evidence linking accumulation of DNA damage to human disease. Similarly, agents that promote cell proliferation (such as tetradecanoylphorbol acetate) often enhance carcinogenesis. For these compounds, the increased likelihood of neoplastic transformation may be a direct consequence of a decrease in the time available for the cell to carry out adequate DNA repair.

Table. Hereditary, cancer-prone disorders that appear to involve defects in DNA repair

Syndrome	Symptoms	Cellular phenotype
Ataxia telangiectasia	Neurological deterioration Immunodeficiency High incidence of lymphoma	Hypersensitivity to ionizing radiation and certain alkylating agents. Dysregulated replication of damaged DNA (may indicate shortened time for DNA repair)
Bloom's syndrome	Developmental abnormalities Lesions on exposed skin High incidence of tumours of the immune system and gastrointestinal tract	High frequency of chromosomal aberrations Defective ligation of breaks associated with DNA repair
Fanconi's anaemia	Growth retardation High incidence of leukaemia	Hypersensitivity to crosslinking agents High frequency of chromosomal aberrations Defective repair of crosslinks in DNA
Hereditary nonpolyposis colon cancer	High incidence of colon cancer	Defect in DNA mismatch repair (when insertion of wrong nucleotide occurs during replication)
Xeroderma pigmentosum	High incidence of epithelioma on exposed areas of skin Neurological impairment (in many cases)	Hypersensitivity to UV light and many chemical carcinogens Defects in excision repair and/or replication of damaged DNA

The earliest theories on how chemicals interact with DNA can be traced back to studies conducted during the development of mustard gas for use in warfare. Further understanding grew out of efforts to identify anticancer agents that would selectively arrest the replication of rapidly dividing tumour cells. Increased public concern over hazards in our environment has prompted additional research into the mechanisms and consequences of chemical interaction with the genetic material. Examples of various types of chemicals which exert genotoxicity are presented in table.

Table. Examples of chemicals that exhibit genotoxicity in human cells

Class of chemical	Example	Source of exposure	Probable genotoxic lesion
Aflatoxins	Aflatoxin B1	Contaminated food	Bulky DNA adducts
Aromatic amines	2-Acetylaminofluorene	Environmental	Bulky DNA adducts
Aziridine quinones	Mitomycin C	Cancer chemotherapy	Mono-adducts, interstrand crosslinks and single-strand breaks in DNA.
Chlorinated hydrocarbons	Vinyl chloride	Environmental	Mono-adducts in DNA
Metals and metal compounds	Cisplatin	Cancer chemotherapy	Both intra- and inter-strand crosslinks in DNA
	Nickel compounds	Environmental	Mono-adducts and single-strand breaks in DNA
Nitrogen mustards	Cyclophosphamide	Cancer chemotherapy	Mono-adducts and interstrand crosslinks in DNA
Nitrosamines	N-Nitrosodimethylamine	Contaminated food	Mono-adducts in DNA
Polycyclic aromatic hydrocarbons	Benzo(a)pyrene	Environmental	Bulky DNA adducts

IMMUNOTOXICOLOGY

The functions of the immune system are to protect the body from invading infectious agents and to provide immune surveillance against arising tumour cells. It has a first line of defence that is non-specific and that can initiate effector reactions itself, and an acquired specific branch, in which lymphocytes and antibodies carry the specificity of recognition and subsequent reactivity towards the antigen.

Immunotoxicology has been defined as "the discipline concerned with the study of the events that can lead to undesired effects as a result of interaction of xenobiotics with the immune system. These undesired events may result as a consequence of (1) a direct and/or indirect effect of the xenobiotic (and/or its biotransformation product) on the immune system, or (2) an immunologically based host response to the compound and/or its metabolite(s), or host antigens modified by the compound or its metabolites".

When the immune system acts as a passive target of chemical insults, the result can be decreased resistance to infection and certain forms of neoplasia, or immune disregulation/stimulation that can exacerbate allergy or auto-immunity.

In the case that the immune system responds to the antigenic specificity of the xenobiotic or host antigen modified by the compound, toxicity can become manifest as allergies or autoimmune diseases.

Animal models to investigate chemical-induced immune suppression have been developed, and a number of these methods are validated. For testing purposes, a tiered approach is followed to make an adequate selection from the overwhelming number of assays available. Generally, the objective of the first tier is to identify potential immunotoxicants. If potential immunotoxicity is identified, a second tier of testing is performed to confirm and characterize further the changes observed. Third-tier investigations include special studies on the mechanism of action of the compound. Several xenobiotics have been identified as immunotoxicants causing immunosuppression in such studies with laboratory animals.

The database on immune function disturbances in humans by environmental chemicals is limited (Descotes 1986; NRC Subcommittee on Immunotoxicology 1992). The use of markers of immunotoxicity has received little attention in clinical and epidemiological studies to investigate the effect of these chemicals on human health. Such studies have not been performed frequently, and their interpretation often does not permit unequivocal conclusions to be drawn, due for instance to the uncontrolled nature of exposure. Therefore, at present, immunotoxicity assessment in rodents, with subsequent extrapolation to man, forms the basis of decisions regarding hazard and risk.

Hypersensitivity reactions, notably allergic asthma and contact dermatitis, are important occupational health problems in industrialized countries. The phenomenon of contact sensitization was investigated first in the guinea pig. Until recently this has been the species of choice for predictive testing. Many guinea pig test methods are available, the most frequently employed being the guinea pig maximization test and the occluded patch test of Buehler. Guinea pig tests and newer approaches developed in mice, such as ear swelling tests and the local lymph node assay, provide the toxicologist with the tools to assess skin sensitization hazard. The situation with respect to sensitization of the respiratory tract is very different. There are, as yet, no well-validated or widely accepted methods available for the identification of chemical respiratory allergens although progress in the development of animal models for the investigation of chemical respiratory allergy has been achieved in the guinea pig and mouse.

Human data show that chemical agents, in particular drugs, can cause autoimmune diseases. There are a number of experimental animal models of human autoimmune diseases. Such comprise both spontaneous pathology (for example systemic lupus erythematosus in New Zealand Black mice) and autoimmune phenomena induced by experimental immunization with a cross-reactive autoantigen (for example the H37Ra adjuvant induced arthritis in Lewis strain rats). These models are applied in the preclinical evaluation of immunosuppressive drugs. Very few studies have addressed the potential of these models for assessment of whether a xenobiotic exacerbates induced or congenital autoimmunity. Animal models that are suitable to investigate the ability of chemicals to induce autoimmune diseases are virtually lacking. One model that is used to a limited extent is the popliteal lymph node assay in mice. Like the situation in humans, genetic factors play a crucial role in the development of autoimmune disease (AD) in laboratory animals, which will limit the predictive value of such tests.

The Immune System

The major function of the immune system is defence against bacteria, viruses, parasites, fungi and neoplastic cells. This is achieved by the actions of various cell types and their soluble mediators in a finely tuned concert. The host defence can be roughly divided into non-specific or innate resistance and specific or acquired immunity mediated by lymphocytes.

Components of the immune system are present throughout the body (Jones *et. al.* 1990). The lymphocyte compartment is found within lymphoid organs. The bone marrow and thymus are classified as primary or central lymphoid organs; the secondary or peripheral lymphoid organs include lymph nodes, spleen and lymphoid tissue along secretory surfaces such as the gastrointestinal and respiratory tracts, the so-called mucosa-associated lymphoid tissue (MALT). About half of the body's lymphocytes are located at any one time in MALT. In addition the skin is an important organ for the induction of immune responses to antigens present on the skin. Important in this process are epidermal Langerhans cells that have an antigen-presenting function.

Phagocytic cells of the monocyte/macrophage lineage, called the mononuclear phagocyte system (MPS), occur in lymphoid organs and also at extranodal sites; the extranodal phagocytes include Kupffer cells in the liver, alveolar macrophages in the lung, mesangial macrophages in the kidney and glial cells in the brain. Polymorphonuclear leukocytes (PMNs) are present mainly in blood and bone marrow, but accumulate at sites of inflammation.

Non-specific Defence

A first line of defence to micro-organisms is executed by a physical and chemical barrier, such as at the skin, the respiratory tract and the alimentary tract. This barrier is helped by non-specific protective mechanisms including phagocytic cells, such as macrophages and polymorphonuclear leukocytes, which are able to kill pathogens, and natural killer cells, which can lyse tumour cells and virus-infected cells. The complement system and certain microbial inhibitors (*e.g.*, lysozyme) also take part in the non-specific response.

Specific Immunity

After initial contact of the host with the pathogen, specific immune responses are induced. The hallmark of this second line of defence is specific recognition of determinants, so-called antigens or epitopes, of the pathogens by receptors on the cell surface of B- and T-lymphocytes. Following interaction with the specific antigen, the receptor-bearing cell is stimulated to undergo proliferation and differentiation, producing a clone of progeny cells that are specific for the eliciting antigen. The specific immune responses help the non-specific defence presented to the pathogens by stimulating the efficacy of the non-specific responses. A fundamental characteristic of specific immunity is that memory develops. Secondary contact with the same antigen provokes a faster and more vigorous but well-regulated response.

The genome does not have the capacity to carry the codes of an array of antigen receptors sufficient to recognize the number of antigens that can be encountered. The repertoire of specificity develops by a process of gene rearrangements. This is a random process, during which various specificities are brought about. This includes specificities for self components, which are undesirable. A selection process that takes place in the thymus (T cells), or bone marrow (B cells) operates to delete these undesirable specificities.

Normal immune effector function and homeostatic regulation of the immune response is dependent upon a variety of soluble products, known collectively as cytokines, which are synthesized and secreted by lymphocytes and by other cell types. Cytokines have pleiotropic effects on immune and inflammatory responses. Cooperation between different cell populations is required for the immune response — the regulation of antibody responses, the accumulation of immune cells and molecules at inflammatory sites, the initiation of acute phase responses, the control of macrophage cytotoxic function and many other processes central to host resistance. These are influenced by, and in many cases are dependent upon, cytokines acting individually or in concert.

Two arms of specific immunity are recognized — humoral immunity and cell-mediated or cellular immunity:

Humoral immunity. In the humoral arm B-lymphocytes are stimulated following recognition of antigen by cell-surface receptors. Antigen receptors on B-lymphocytes are immunoglobulins (Ig). Mature B cells (plasma cells) start the production of antigen-specific immunoglobulins that act as antibodies in serum or along mucosal surfaces. There are five major classes of immunoglobulins: (1) IgM, pentameric Ig with optimal agglutinating capacity, which is first produced after antigenic stimulation; (2) IgG, the main Ig in circulation, which can pass the placenta; (3) IgA, secretory Ig for the protection of mucosal surfaces; (4) IgE, Ig fixing to mast cells or basophilic granulocytes involved in immediate hypersensitivity reactions and (5) IgD, whose major function is as a receptor on B-lymphocytes.

Cell-mediated immunity. The cellular arm of the specific immune system is mediated by T-lymphocytes. These cells also have antigen receptors on their membranes. They recognize antigen if presented by antigen presenting cells in the context of histocompatibility antigens. Hence, these cells have a restriction in addition to the antigen specificity. T cells function as helper cells for various (including humoral) immune responses, mediate recruitment of inflammatory cells, and can, as cytotoxic T cells, kill target cells after antigen-specific recognition.

MECHANISMS OF IMMUNOTOXICITY

Immunosuppression

Effective host resistance is dependent upon the functional integrity of the immune system, which in turn requires that the component cells and molecules which orchestrate immune responses are available in sufficient numbers and in an operational form. Congenital immunodeficiencies in humans are often characterized by defects in certain stem cell lines, resulting in impaired or absent production of immune cells. By analogy with congenital and acquired human immunodeficiency diseases, chemical-induced immunosuppression may result simply from a reduced number of functional cells (IPCS 1996). The absence, or reduced numbers, of lymphocytes may have more or less profound effects on immune status. Some immunodeficiency states and severe immunosuppression, as can occur in transplantation or cytostatic therapy, have been associated in particular with increased incidences of opportunistic infections and of certain neoplastic diseases. The infections can be bacterial, viral, fungal or protozoan, and the predominant type of infection depends on the associated immunodeficiency. Exposure to immunosuppressive environmental chemicals may be expected to result in more subtle forms of immunosuppression, which may be difficult to detect. These may lead, for example, to an increased incidence of infections such as influenza or the common cold.

In view of the complexity of the immune system, with the wide variety of cells, mediators and functions that form a complicated and interactive network, immunotoxic compounds have numerous opportunities to exert an effect. Although the nature of the initial lesions induced by many immunotoxic chemicals have not yet been elucidated, there is increasing information available, mostly derived from studies in laboratory animals, regarding the immunobiological changes which result in depression of immune function. Toxic effects might occur at the following critical functions (and some examples are given of immunotoxic compounds affecting these functions):

- development and expansion of different stem cell populations (benzene exerts immunotoxic effects at the stem cell level, causing lymphocytopenia)
- proliferation of various lymphoid and myeloid cells as well as supportive tissues in which these cells mature and function (immunotoxic organotin compounds suppress the proliferative activity of lymphocytes in the thymic cortex through direct cytotoxicity; the thymotoxic action of 2,3,7,8-tetrachloro-dibenzo-p-dioxin (TCDD) and related compounds is likely due to an impaired function of thymic epithelial cells, rather than to direct toxicity for thymocytes)
- antigen uptake, processing and presentation by macrophages and other antigen-presenting cells (one of the targets of 7,12-dimethylbenz(a)anthracene (DMBA) and of lead is antigen presentation by macrophages; a target of ultraviolet radiation is the antigen-presenting Langerhans cell)

- regulatory function of T-helper and T-suppressor cells (T-helper cell function is impaired by organotins, aldicarb, polychlorinated biphenyls (PCBs), TCDD and DMBA; T-suppressor cell function is reduced by low-dose cyclophosphamide treatment)

- production of various cytokines or interleukins (benzo(a)pyrene (BP) suppresses interleukin-1 production; ultraviolet radiation alters production of cytokines by keratinocytes)

- synthesis of various classes of immunoglobulins IgM and IgG is suppressed following PCB and tributyltin oxide (TBT) treatment, and increased after hexachlorobenzene (HCB) exposure).

- complement regulation and activation (affected by TCDD)

- cytotoxic T cell function (3-methylcholanthrene (3-MC), DMBA, and TCDD suppress cytotoxic T cell activity)

- natural killer (NK) cell function (pulmonary NK activity is suppressed by ozone; splenic NK activity is impaired by nickel)

- macrophage and polymorphonuclear leukocyte chemotaxis and cytotoxic functions (ozone and nitrogen dioxide impair the phagocytic activity of alveolar macrophages).

Allergy

Allergy may be defined as the adverse health effects which result from the induction and elicitation of specific immune responses. When hypersensitivity reactions occur without involvement of the immune system the term pseudo-allergy is used. In the context of immunotoxicology, allergy results from a specific immune response to chemicals and drugs that are of interest. The ability of a chemical to sensitize individuals is generally related to its ability to bind covalently to body proteins. Allergic reactions may take a variety of forms and these differ with respect to both the underlying immunological mechanisms and the speed of the reaction. Four major types of allergic reactions have been recognized: Type I hypersensitivity reactions, which are effectuated by IgE antibody and where symptoms are manifest within minutes of exposure of the sensitized individual. Type II hypersensitivity reactions result from the damage or destruction of host cells by antibody. In this case symptoms become apparent within hours. Type III hypersensitivity, or Arthus, reactions are also antibody mediated, but against soluble antigen, and result from the local or systemic action of immune complexes. Type IV, or delayed-type hypersensitivity, reactions are effected by T-lymphocytes and normally symptoms develop 24 to 48 hours following exposure of the sensitized individual.

The two types of chemical allergy of greatest relevance to occupational health are contact sensitivity or skin allergy and allergy of the respiratory tract.

Contact hypersensitivity. A large number of chemicals are able to cause skin sensitization. Following topical exposure of a susceptible individual to a chemical allergen, a T-lymphocyte response is induced in the draining lymph nodes. In the skin the allergen interacts directly or indirectly with epidermal Langerhans cells, which transport the chemical to the lymph nodes and present it in an immunogenic form to responsive T-lymphocytes. Allergen-activated T-lymphocytes proliferate, resulting in clonal expansion. The individual is now sensitized and will respond to a second dermal exposure to the same chemical with a more aggressive immune response, resulting in allergic contact dermatitis. The cutaneous inflammatory reaction which characterizes allergic contact dermatitis is secondary to the recognition of the allergen in the skin by specific T-lymphocytes. These lymphocytes become activated, release cytokines and cause the local accumulation of other mononuclear leukocytes. Symptoms develop some 24 to 48 hours following exposure of the sensitized individual, and allergic contact dermatitis therefore represents a form of delayed-type hypersensitivity. Common causes of allergic contact dermatitis include organic chemicals (such as 2,4-dinitrochlorobenzene), metals (such as nickel and chromium) and plant products (such as urushiol from poison ivy).

Respiratory hypersensitivity. Respiratory hypersensitivity is usually considered to be a Type I hypersensitivity reaction. However, late phase reactions and the more chronic symptoms associated with asthma may involve cell-mediated (Type IV) immune processes. The acute symptoms associated with respiratory allergy are effected by IgE antibody, the production of which is provoked following exposure of the susceptible individual to the inducing chemical allergen. The IgE antibody distributes systemically and binds, *via* membrane receptors, to mast cells which are found in vascularized tissues, including the respiratory tract. Following inhalation of the same chemical a respiratory hypersensitivity reaction will be elicited. Allergen associates with protein and binds to, and cross-links, IgE antibody bound to mast cells. This in turn causes the degranulation of mast cells and the release of inflammatory mediators such as histamine and leukotrienes. Such mediators cause bronchoconstriction and vasodilation, resulting in the symptoms of respiratory allergy; asthma and/or rhinitis. Chemicals known to cause respiratory hypersensitivity in man include acid anhydrides (such as trimellitic anhydride), some diisocyanates (such as toluene diisocyanate), platinum salts and some reactive dyes. Also, chronic exposure to beryllium is known to cause hypersensitivity lung disease.

Autoimmunity

Autoimmunity can be defined as the stimulation of specific immune responses directed against endogenous "self" antigens. Induced autoimmunity can result either from alterations in the balance of regulatory T-lymphocytes or from the association of a xenobiotic with normal tissue components such as to render them immunogenic ("altered self"). Drugs and chemicals known to incidentally induce or exacerbate effects like those of autoimmune disease (AD) in susceptible individuals are low molecular weight compounds (molecular weight 100 to 500) that are generally considered to be not immunogenic themselves. The mechanism of AD by chemical exposure is mostly unknown. Disease can be produced directly by means of circulating antibody, indirectly through the formation of immune complexes, or as a consequence of cell-mediated immunity, but likely occurs through a combination of mechanisms. The pathogenesis is best known in immune haemolytic disorders induced by drugs:

- The drug can attach to the red-cell membrane and interact with a drug-specific antibody.

- The drug can alter the red-cell membrane so that the immune system regards the cell as foreign.

- The drug and its specific antibody form immune complexes that adhere to the red-cell membrane to produce injury.

- Red-cell sensitization occurs due to the production of red-cell autoantibody.

A variety of chemicals and drugs, in particular the latter, have been found to induce autoimmune-like responses (Kamüller, Bloksma and Seinen 1989). Occupational exposure to chemicals may incidentally lead to AD-like syndromes. Exposure to monomeric vinyl chloride, trichloroethylene, perchloroethylene, epoxy resins and silica dust may induce scleroderma-like syndromes. A syndrome similar to systemic lupus erythematosus (SLE) has been described after exposure to hydrazine. Exposure to toluene diisocyanate has been associated with the induction of thrombocytopenic purpura. Heavy metals such as mercury have been implicated in some cases of immune complex glomerulonephritis.

Human Risk Assessment

The assessment of human immune status is performed mainly using peripheral blood for analysis of humoral substances like immunoglobulins and complement, and of blood leukocytes for subset composition and functionality of subpopulations. These methods are usually the same as those used to investigate humoral and cell-mediated immunity as well as nonspecific resistance of patients with suspected congenital immunodeficiency disease. For epidemiological studies (*e.g.*, of occupationally exposed populations) parameters should be selected on the basis of their predictive value in human populations, validated animal models, and the underlying biology of the markers.

The strategy in screening for immunotoxic effects after (accidental) exposure to environmental pollutants or other toxicants is much dependent on circumstances, such as type of immunodeficiency to be expected, time between exposure and immune status assessment, degree of exposure and number of exposed individuals. The process of assessing the immunotoxic risk of a particular xenobiotic in humans is extremely difficult and often impossible, due largely to the presence of various confounding factors of endogenous or exogenous origin that influence the response of individuals to toxic damage. This is particularly true for studies which investigate the role of chemical exposure in autoimmune diseases, where genetic factors play a crucial role.

Table. Classification of tests for immune markers

Test category	Characteristics	Specific tests
Basic-general Should be included with general panels	Indicators of general health and organ system status	Blood urea nitrogen, blood glucose, *etc.*
Basic-immune Should be included with general panels	General indicators of immune status Relatively low cost Assay methods are standardized among laboratories Results outside reference ranges are clinically interpretable	Complete blood counts Serum IgG, IgA, IgM levels Surface marker phenotypes for major lymphocyte subsets
Focused/reflex Should be included when indicated by clinical findings, suspected exposures, or prior test results	Indicators of specific immune functions/events Cost varies Assay methods are standardized among laboratories Results outside reference ranges are clinically interpretable	Histocompatibility genotype Antibodies to infectious agents Total serum IgE Allergen-specific IgE Autoantibodies Skin tests for hypersensitivity Granulocyte oxidative burst Histopathology (tissue biopsy)
Research Should be included only with control populations and careful study design	Indicators of general or specific immune functions/events Cost varies; often expensive Assay methods are usually not standardized among laboratories Results outside reference ranges are often not clinically interpretable	In vitro stimulation assays Cell activation surface markers Cytokine serum concentrations Clonality assays (antibody, cellular, genetic) Cytotoxicity tests

As adequate human data are seldom available, the assessment of risk for chemical-induced immunosuppression in humans is in the majority of cases based upon animal studies. The identification of potential immunotoxic xenobiotics is undertaken primarily in controlled studies in rodents. In vivo exposure studies present, in this regard, the optimal approach to estimate the immunotoxic potential of a compound. This is due to the multifactoral and complex nature of the immune system and of immune responses. In vitro studies are of increasing value in the elucidation of mechanisms of immunotoxicity. In addition, by investigating the effects of the compound using cells of animal and human origin, data can be generated for species comparison, which can be used in the "parallelogram" approach to improve the risk assessment process. If data are available for three cornerstones of the parallelogram (in vivo animal, and in vitro animal and human) it may be easier to predict the outcome at the remaining cornerstone, that is, the risk in humans.

When assessment of risk for chemical-induced immunosuppression has to rely solely upon data from animal studies, an approach can be followed in the extrapolation to man by the application of uncertainty factors to the no observed adverse effect level (NOAEL). This level can be based on parameters determined in relevant models, such as host resistance assays and in vivo assessment of hypersensitivity reactions and antibody production. Ideally, the relevance of this approach to risk assessment requires confirmation by studies in humans. Such studies should combine the identification and measurement of the toxicant, epidemiological data and immune status assessments.

To predict contact hypersensitivity, guinea pig models are available and have been used in risk assessment since the 1970s. Although sensitive and reproducible, these tests have limitations as they depend on subjective evaluation; this can be overcome by newer and more quantitative methods developed in the mouse. Regarding chemical-induced hypersensitivity induced by inhalation or ingestion of allergens, tests should be developed and evaluated in terms of their predictive value in man. When it comes to setting safe occupational exposure levels of potential allergens, consideration has to be given to the biphasic nature of allergy: the sensitization phase and the elicitation phase. The concentration required to elicit an allergic reaction in a previously sensitized individual is considerably lower than the concentration necessary to induce sensitization in the immunologically naïve but susceptible individual.

As animal models to predict chemical-induced autoimmunity are virtually lacking, emphasis should be given to the development of such models. For the development of such models, our knowledge of chemical-induced autoimmunity in humans should be advanced, including the study of genetic and immune system markers to identify susceptible individuals. Humans that are exposed to drugs that induce autoimmunity offer such an opportunity.

TARGET ORGAN TOXICOLOGY

The study and characterization of chemicals and other agents for toxic properties is often undertaken on the basis of specific organs and organ systems. In this chapter, two targets have been selected for in-depth discussion: the immune system and the gene. These examples were chosen to represent a complex target organ system and a molecular target within cells. For more comprehensive discussion of the toxicology of target organs, the reader is referred to standard toxicology texts such as Casarett and Doull, and Hayes. The International Programme on Chemical Safety (IPCS) has also published several criteria documents on target organ toxicology, by organ system.

Target organ toxicology studies are usually undertaken on the basis of information indicating the potential for specific toxic effects of a substance, either from epidemiological data or from general acute or chronic toxicity studies, or on the basis of special concerns to protect certain organ functions, such as reproduction or foetal development. In some cases, specific target organ toxicity tests are expressly mandated by statutory authorities, such as neurotoxicity testing under the US pesticides law and mutagenicity testing under the Japanese Chemical Substance Control Law.

As discussed in "Target organ and critical effects," the identification of a critical organ is based upon the detection of the organ or organ system which first responds adversely or to the lowest doses or exposures. This information is then used to design specific toxicology investigations or more defined toxicity tests that are designed to elicit more sensitive indications of intoxication in the target organ. Target organ toxicology studies may also be used to determine mechanisms of action, of use in risk assessment.

Methods of Target Organ Toxicity Studies

Target organs may be studied by exposure of intact organisms and detailed analysis of function and histopathology in the target organ, or by in vitro exposure of cells, tissue slices, or whole organs maintained for short or long term periods in culture (see "Mechanisms of toxicology: Introduction and concepts"). In some cases, tissues from human subjects may also be available for target organ toxicity studies, and these may provide opportunities to validate assumptions of cross-species extrapolation. However, it must be kept in mind that such studies do not provide information on relative toxicokinetics.

In general, target organ toxicity studies share the following common characteristics: detailed histopathological examination of the target organ, including post mortem examination, tissue weight, and examination of fixed tissues; biochemical studies of critical pathways in the target organ, such as important enzyme systems; functional studies of the ability of the organ and cellular constituents to perform expected metabolic and other functions; and analysis of biomarkers of exposure and early effects in target organ cells.

Detailed knowledge of target organ physiology, biochemistry and molecular biology may be incorporated in target organ studies. For instance, because the synthesis and secretion of small-molecular-weight proteins is an important aspect of renal function, nephrotoxicity studies often include special attention to these parameters (IPCS 1991). Because cell-to-cell communication is a fundamental process of nervous system function, target organ studies in neurotoxicity may include detailed neurochemical and biophysical measurements of neurotransmitter synthesis, uptake, storage, release and receptor binding, as well as electrophysiological measurement of changes in membrane potential associated with these events.

A high degree of emphasis is being placed upon the development of in vitro methods for target organ toxicity, to replace or reduce the use of whole animals. Substantial advances in these methods have been achieved for reproductive toxicants (Heindel and Chapin 1993).

In summary, target organ toxicity studies are generally undertaken as a higher order test for determining toxicity. The selection of specific target organs for further evaluation depends upon the results of screening level tests, such as the acute or subchronic tests used by OECD and the European Union; some target organs and organ systems may be a priori candidates for special investigation because of concerns to prevent certain types of adverse health effects.

Biomarkers

The word biomarker is short for biological marker, a term that refers to a measurable event occurring in a biological system, such as the human body. This event is then interpreted as a reflection, or marker, of a more general state of the organism or of life expectancy. In occupational health, a biomarker is generally used as an indicator of health status or disease risk.

Biomarkers are used for in vitro as well as in vivo studies that may include humans. Usually, three specific types of biological markers are identified. Although a few biomarkers may be difficult to classify, usually they are separated into biomarkers of exposure, biomarkers of effect or biomarkers of susceptibility.

Table. Examples of biomarkers of exposure or biomarkers of effect that are used in toxicological studies in occupational health

Sample	Measurement	Purpose
Exposure biomarkers		
Adipose tissue	Dioxin	Dioxin exposure
Blood	Lead	Lead exposure
Bone	Aluminium	Aluminium exposure
Exhaled breath	Toluene	Toluene exposure
Hair	Mercury	Methylmercury exposure
Serum	Benzene	Benzene exposure
Urine	Phenol	Benzene exposure
Effect biomarkers		
Blood	Carboxyhaemoglobin	Carbon monoxide exposure
Red blood cells	Zinc-protoporphyrin	Lead exposure
Serum	Cholinesterase	Organophosphate exposure
Urine	Microglobulins	Nephrotoxic exposure
White blood cells	DNA adducts	Mutagen exposure

Given an acceptable degree of validity, biomarkers may be employed for several purposes. On an individual basis, a biomarker may be used to support or refute a diagnosis of a particular type of poisoning or other chemically-induced adverse effect. In a healthy subject, a biomarker may also reflect individual hypersusceptibility to specific chemical exposures and may therefore serve as a basis for risk prediction and counselling. In groups of exposed workers, some exposure biomarkers can be applied to assess the extent of compliance with pollution abatement regulations or the effectiveness of preventive efforts in general.

Biomarkers of Exposure

An exposure biomarker may be an exogenous compound (or a metabolite) within the body, an interactive product between the compound (or metabolite) and an endogenous component, or another event related to the exposure. Most commonly, biomarkers of exposures to stable compounds, such as metals, comprise measurements of the metal concentrations in appropriate samples, such as blood, serum or urine. With volatile chemicals, their concentration in exhaled breath (after inhalation of contamination-free air) may be assessed. If the compound is metabolized in the body, one or more metabolites may be chosen as a biomarker of the exposure; metabolites are often determined in urine samples.

Modern methods of analysis may allow separation of isomers or congeners of organic compounds, and determination of the speciation of metal compounds or isotopic ratios of certain elements. Sophisticated analyses allow determination of changes in the structure of DNA or other macromolecules caused by binding with reactive chemicals. Such advanced techniques will no doubt gain considerably in importance for applications in biomarker studies, and lower detection limits and better analytical validity are likely to make these biomarkers even more useful.

Particularly promising developments have occurred with biomarkers of exposure to mutagenic chemicals. These compounds are reactive and may form adducts with macromolecules, such as proteins or DNA. DNA adducts may be detected in white blood cells or tissue biopsies, and specific DNA fragments may be excreted in the urine. For example, exposure to ethylene oxide results in reactions with DNA bases, and, after excision of the damaged base, N-7-(2-hydroxyethyl) guanine will be eliminated in the urine. Some adducts may not refer directly to a particular exposure. For example, 8-hydroxy-2′-deoxyguanosine reflects oxidative damage to DNA, and this reaction may be triggered by several chemical compounds, most of which also induce lipid peroxidation.

Other macromolecules may also be changed by adduct formation or oxidation. Of special interest, such reactive compounds may generate haemoglobin adducts that can be determined as biomarkers of exposure to the compounds. The advantage is that ample amounts of haemoglobin can be obtained from a blood sample, and, given the four-month lifetime of red blood cells, the adducts formed with the amino acids of the protein will indicate the total exposure during this period.

Adducts may be determined by sensitive techniques such as high-performance lipid chromatography, and some immunological methods are also available. In general, the analytical methods are new, expensive and need further development and validation. Better sensitivity can be obtained by using the ^{32}P post labelling assay, which is a nonspecific indication that DNA damage has taken place. All of these techniques are potentially useful for biological monitoring and have been applied in a growing number of studies. However, simpler and more sensitive analytical methods are needed. Given the limited specificity of some methods at low-level exposures, tobacco smoking or other factors may impact significantly on the measurement results, thus causing difficulties in interpretation.

Exposure to mutagenic compounds, or to compounds which are metabolized into mutagens, may also be determined by assessing the mutagenicity of the urine from an exposed individual. The urine sample is incubated with a strain of bacteria in which a specific point mutation is expressed in a way that can be easily measured. If mutagenic chemicals are present in the urine sample, then an increased rate of mutations will occur in the bacteria.

Exposure biomarkers must be evaluated with regard to temporal variation in exposure and the relation to different compartments. Thus, the time frame(s) represented by the biomarker, that is, the extent to which the biomarker measurement reflects past exposure(s) and/or accumulated body burden, must be determined from toxicokinetic data in order to interpret the result. In particular, the degree to which the biomarker indicates retention in specific target organs should be considered. Although blood samples are often used for biomarker studies, peripheral blood is generally not regarded as a compartment as such, although it acts as a transport medium between compartments. The degree to which the concentration in the blood reflects levels in different organs varies widely between different chemicals, and usually also depends upon the length of the exposure as well as time since exposure.

Sometimes this type of evidence is used to classify a biomarker as an indicator of (total) absorbed dose or an indicator of effective dose (*i.e.*, the amount that has reached the target tissue). For example, exposure to a particular solvent may be evaluated from data on the actual concentration of the solvent in the blood at a particular time following the exposure. This measurement will reflect the amount of the solvent that has been absorbed into the body. Some of the absorbed amount will be exhaled due to the vapour pressure of the solvent. While circulating in the blood, the solvent will interact with various components of the body, and it will eventually become subject to breakdown by enzymes. The outcome of the metabolic processes can be assessed by determining specific mercapturic acids produced by conjugation with glutathione. The cumulative excretion of mercapturic acids may better reflect the effective dose than will the blood concentration.

Life events, such as reproduction and senescence, may affect the distribution of a chemical. The distribution of chemicals within the body is significantly affected by pregnancy, and many chemicals may pass the placental barrier, thus causing exposure of the foetus. Lactation may result in excretion of lipid-soluble chemicals, thus leading to a decreased retention in the mother along with an increased uptake by the infant. During weight loss or development of osteoporosis, stored chemicals may be released, which can then result in a renewed and protracted "endogenous" exposure of target organs. Other factors may affect individual absorption, metabolism, retention and distribution of chemical compounds, and some biomarkers of susceptibility are available.

Biomarkers of Effect

A marker of effect may be an endogenous component, or a measure of the functional capacity, or some other indicator of the state or balance of the body or organ system, as affected by the exposure. Such effect markers are generally preclinical indicators of abnormalities.

These biomarkers may be specific or non-specific. The specific biomarkers are useful because they indicate a biological effect of a particular exposure, thus providing evidence that can potentially be used for preventive purposes. The non-specific biomarkers do not point to an individual cause of the effect, but they may reflect the total, integrated effect due to a mixed exposure. Both types of biomarkers may therefore be of considerable use in occupational health.

There is not a clear distinction between exposure biomarkers and effect bio-markers. For example, adduct formation could be said to reflect an effect rather than the exposure. However, effect biomarkers usually indicate changes in the functions of cells, tissues or the total body. Some researchers include gross changes, such as an increase in liver weight of exposed laboratory animals or decreased growth in children, as biomarkers of effect. For the purpose of occupational health, effect biomarkers should be restricted to those that indicate subclinical or revers-ible biochemical changes, such as inhibition of enzymes. The most frequently used effect biomarker is probably inhibition of cholinesterase caused by certain insecticides, that is, organophosphates and carbamates. In most cases, this effect is entirely reversible, and the enzyme inhibition reflects the total exposure to this particular group of insecticides.

Some exposures do not result in enzyme inhibition but rather in increased activity of an enzyme. This is the case with several enzymes that belong to the P450 family (see "Genetic determinants of toxic response"). They may be induced by exposures to certain solvents and polyaromatic hydrocarbons (PAHs). Since these enzymes are mainly expressed in tissues from which a biopsy may be difficult to obtain, the enzyme activity is determined indirectly in vivo by administering a compound that is metabolized by that particular enzyme, and then the breakdown product is measured in urine or plasma.

Other exposures may induce the synthesis of a protective protein in the body. The best example is probably metallothionein, which binds cadmium and promotes the excretion of this metal; cadmium exposure is one of the factors that result in increased expression of the metallothionein gene. Similar protective proteins may exist but have not yet been explored sufficiently to become accepted as biomarkers. Among the candidates for possible use as biomarkers are the so-called stress proteins, originally referred to as heat shock proteins. These proteins are generated by a range of different organisms in response to a variety of adverse exposures.

Oxidative damage may be assessed by determining the concentration of malondialdehyde in serum or the exhalation of ethane. Similarly, the urinary excretion of proteins with a small molecular weight, such as albumin, may be used as a biomarker of early kidney damage. Several parameters routinely used in clinical practice (for example, serum hormone or enzyme levels) may also be useful as biomarkers. However, many of these parameters may not be sufficiently sensitive to detect early impairment.

Another group of effect parameters relate to genotoxic effects (changes in the structure of chromosomes). Such effects may be detected by microscopy of white blood cells that undergo cell division. Serious damage to the chromosomes — chromosomal aberrations or formation of micronuclei — can be seen in a microscope. Damage may also be revealed by adding a dye to the cells during cell division. Exposure to a genotoxic agent can then be visualized as an increased exchange of the dye between the two chromatids of each chromosome (sister chromatid exchange). Chromosomal aberrations are related to an increased risk of developing cancer, but the significance of an increased rate of sister chromatid exchange is less clear.

More sophisticated assessment of genotoxicity is based on particular point mutations in somatic cells, that is, white blood cells or epithelial cells obtained from the oral mucosa. A mutation at a specific locus may make the cells capable of growing in a culture that contains a chemical that is otherwise toxic (such as 6-thioguanine). Alternatively, a specific gene product can be assessed (e.g., serum or tissue concentrations of oncoproteins encoded by particular oncogenes). Obviously, these mutations reflect the total genotoxic damage incurred and do not necessarily indicate anything about the causative exposure. These methods are not yet ready for practical use in occupational health, but rapid progress in this line of research would suggest that such methods will become available within a few years.

Biomarkers of Susceptibility

A marker of susceptibility, whether inherited or induced, is an indicator that the individual is particularly sensitive to the effect of a xenobiotic or to the effects of a group of such compounds. Most attention has been focused on genetic susceptibility, although other factors may be at least as important. Hypersusceptibility may be due to an inherited trait, the constitution of the individual, or environmental factors.

The ability to metabolize certain chemicals is variable and is genetically determined (see "Genetic determinants of toxic response"). Several relevant enzymes appear to be controlled by a single gene. For example, oxidation of foreign chemicals is mainly carried out be a family of enzymes belonging to the P450 family. Other enzymes make the metabolites more water soluble by conjugation (*e.g.*, N-acetyltransferase and μ-glutathion-S-transferase). The activity of these enzymes is genetically controlled and varies considerably. As mentioned above, the activity can be determined by administering a small dose of a drug and then determining the amount of the metabolite in the urine. Some of the genes have now been characterized, and techniques are available to determine the genotype. Important studies suggest that a risk of developing certain cancer forms is related to the capability of metabolizing foreign compounds. Many questions still remain unanswered, thus at this time limiting the use of these potential susceptibility biomarkers in occupational health.

Other inherited traits, such as alpha$_1$-antitrypsin deficiency or glucose-6-phosphate dehydrogenase deficiency, also result in deficient defence mechanisms in the body, thereby causing hypersusceptibility to certain exposures.

Most research related to susceptibility has dealt with genetic predisposition. Other factors play a role as well and have been partly neglected. For example, individuals with a chronic disease may be more sensitive to an occupational exposure. Also, if a disease process or previous exposure to toxic chemicals has caused some subclinical organ damage, then the capacity to withstand a new toxic exposure is likely to be less. Biochemical indicators of organ function may in this case be used as susceptibility biomarkers. Perhaps the best example regarding hypersusceptibility relates to allergic responses. If an individual has become sensitized to a particular exposure, then specific antibodies can be detected in serum. Even if the individual has not become sensitized, other current or past exposures may add to the risk of developing an adverse effect related to an occupational exposure.

A major problem is to determine the joint effect of mixed exposures at work. In addition, personal habits and drug use may result in an increased susceptibility. For example, tobacco smoke usually contains a considerable amount of cadmium. Thus, with occupational exposure to cadmium, a heavy smoker who has accumulated substantial amounts of this metal in the body will be at increased risk of developing cadmium-related kidney disease.

Application in Occupational Health

Biomarkers are extremely useful in toxicological research, and many may be applicable in biological monitoring. Nonetheless, the limitations must also be recognized. Many biomarkers have so far been studied only in laboratory animals. Toxicokinetic patterns in other species may not necessarily reflect the situation in human beings, and extrapolation may require confirmatory studies in human volunteers. Also, account must be taken of individual variations due to genetic or constitutional factors.

In some cases, exposure biomarkers may not at all be feasible (*e.g.*, for chemicals which are short-lived in vivo). Other chemicals may be stored in, or may affect, organs which cannot be accessed by routine procedures, such as the nervous system. The route of exposure may also affect the distribution pattern and therefore also the biomarker measurement and its interpretation. For example, direct exposure of the brain *via* the olfactory nerve is likely to escape detection by measurement of exposure biomarkers. As to effect biomarkers, many of them are not at all specific, and the change can be due to a variety of causes, including lifestyle factors. Perhaps in particular with the susceptibility biomarkers, interpretation must be very cautious at the moment, as many uncertainties remain about the overall health significance of individual genotypes.

In occupational health, the ideal biomarker should satisfy several requirements. First of all, sample collection and analysis must be simple and reliable. For optimal analytical quality, standardization is needed, but the specific requirements vary considerably. Major areas of concern include: preparation of the individual, sampling procedure and sample handling, and measurement procedure; the latter encompasses technical factors, such as calibration and quality assurance procedures, and individual-related factors, such as education and training of operators.

For documentation of analytical validity and traceability, reference materials should be based on relevant matrices and with appropriate concentrations of toxic substances or relevant metabolites at appropriate levels. For biomarkers to be used for biological monitoring or for diagnostic purposes, the responsible laboratories must have well-documented analytical procedures with defined performance characteristics, and accessible records to allow verification of the results. At the same time, nonetheless, the economics of characterizing and using reference materials to supplement quality assurance procedures in general must be considered. Thus, the achievable quality of results, and the uses to which they are put, have to be balanced against the added costs of quality assurance, including reference materials, manpower and instrumentation.

Another requirement is that the biomarker should be specific, at least under the circumstances of the study, for a particular type of exposure, with a clear-cut relationship to the degree of exposure. Otherwise, the result of the biomarker measurement may be too difficult to interpret. For proper interpretation of the measurement result of an exposure biomarker, the diagnostic validity must be known (*i.e.*, the translation of the biomarker value into the magnitude of possible health risks). In this area, metals serve as a paradigm for biomarker research. Recent research has demonstrated the complexity and subtlety of dose-response relationships, with considerable difficulty in identifying no-effect levels and therefore also in defining tolerable exposures. However, this kind of research has also illustrated the types of investigation and the refinement that are necessary to uncover the relevant information. For most organic compounds, quantitative associations between exposures and the corresponding adverse health effects are not yet available; in many cases, even the primary target organs are not known for sure. In addition, evaluation of toxicity data and biomarker concentrations is often complicated by exposure to mixtures of substances, rather than exposure to a single compound at the time.

Before the biomarker is applied for occupational health purposes, some additional considerations are necessary. First, the biomarker must reflect a subclinical and reversible change only. Second, given that the biomarker results can be interpreted with regard to health risks, then preventive efforts should be available and should be considered realistic in case the biomarker data suggests a need to reduce the exposure. Third, the practical use of the biomarker must be generally regarded as ethically acceptable.

Industrial hygiene measurements may be compared with applicable exposure limits. Likewise, results on exposure biomarkers or effect biomarkers may be compared to biological action limits, sometimes referred to as biological exposure indices. Such limits should be based on the best advice of clinicians and scientists from appropriate disciplines, and responsible administrators as "risk managers" should then take into account relevant ethical, social, cultural and economic factors. The scientific basis should, if possible, include dose-response relationships supplemented by information on variations in susceptibility within the population at risk. In some countries, workers and members of the general public are involved in the standard-setting process and provide important input, particularly when scientific uncertainty is considerable. One of the major uncertainties is how to define an adverse health effect that should be prevented — for example, whether adduct formation as an exposure biomarker by itself represents an adverse effect (*i.e.*, effect biomarker) that should be prevented. Difficult questions are likely to arise when deciding whether it is ethically defensible, for the same compound, to have different limits for adventitious exposure, on the one hand, and occupational exposure, on the other.

The information generated by the use of biomarkers should generally be conveyed to the individuals examined within the physician-patient relationship. Ethical concerns must in particular be considered in connection with highly experimental biomarker analyses that cannot currently be interpreted in detail in terms of actual health risks. For the general population, for example, limited guidance exists at present with regard to interpretation of exposure biomarkers other than the blood-lead concentration. Also of importance is the confidence in the data generated (*i.e.*, whether appropriate sampling has been done, and whether sound quality assurance procedures have been utilized in the laboratory involved). An additional area of special worry relates to individual hypersusceptibility. These issues must be taken into account when providing the feedback from the study.

All sectors of society affected by, or concerned with carrying out, a biomarker study need to be involved in the decision-making process on how to handle the information generated by the study. Specific procedures to prevent or overcome inevitable ethical conflicts should be developed within the legal and social frameworks of the region or country. However, each situation represents a different set of questions and pitfalls, and no single procedure for public involvement can be developed to cover all applications of exposure biomarkers.

GENETIC TOXICITY ASSESSMENT

Genetic toxicity assessment is the evaluation of agents for their ability to induce any of three general types of changes (mutations) in the genetic material

(DNA): gene, chromosomal and genomic. In organisms such as humans, the genes are composed of DNA, which consists of individual units called nucleotide bases. The genes are arranged in discrete physical structures called chromosomes. Genotoxicity can result in significant and irreversible effects upon human health. Genotoxic damage is a critical step in the induction of cancer and it can also be involved in the induction of birth defects and foetal death. The three classes of mutations mentioned above can occur within either of the two types of tissues possessed by organisms such as humans: sperm or eggs (germ cells) and the remaining tissue (somatic cells).

Assays that measure gene mutation are those that detect the substitution, addition or deletion of nucleotides within a gene. Assays that measure chromosomal mutation are those that detect breaks or chromosomal rearrangements involving one or more chromosomes. Assays that measure genomic mutation are those that detect changes in the number of chromosomes, a condition called aneuploidy. Genetic toxicity assessment has changed considerably since the development by Herman Muller in 1927 of the first assay to detect genotoxic (mutagenic) agents. Since then, more than 200 assays have been developed that measure mutations in DNA; however, fewer than ten assays are used commonly today for genetic toxicity assessment. This article reviews these assays, describes what they measure, and explores the role of these assays in toxicity assessment.

Identification of Cancer Hazards Prior to the Development of the Field of Genetic Toxicology

Genetic toxicology has become an integral part of the overall risk assessment process and has gained in stature in recent times as a reliable predictor for carcinogenic activity. However, prior to the development of genetic toxicology (before 1970), other methods were and are still being used to identify potential cancer hazards to humans. There are six major categories of methods currently used for identifying human cancer risks: epidemiological studies, long-term in vivo bioassays, mid-term in vivo bioassays, short-term in vivo and in vitro bioassays, artificial intelligence (structure-activity), and mechanism-based inference.

Table gives advantages and disadvantages for these methods.

Table. Advantages and disadvantages of current methods for identifying human cancer risks

	Advantages	Disadvantages
Epidemiological studies	(1) humans are ultimate indicators of disease; (2) evaluate sensitive or susceptible populations; (3) occupational exposure cohorts; (4) environmental sentinel alerts	(1) generally retrospective (death certificates, recall biases, *etc.*); (2) insensitive, costly, lengthy; (3) reliable exposure data sometimes unavailable or difficult to obtain; (4) combined, multiple and complex exposures; lack of appropriate control cohorts; (5) experiments on humans not done; (6) cancer detection, not prevention

Long-term in vivo bioassays	(1) prospective and retrospective (validation) evaluations; (2) excellent correlation with identified human carcinogens; (3) exposure levels and conditions known; (4) identifies chemical toxicity and carcinogenicity effects; (5) results obtained relatively quickly; (6) qualitative comparisons among chemical classes; (7) integrative and interactive biologic systems related closely to humans	(1) rarely replicated, resource intensive; (3) limited facilities suitable for such experiments; (4) species extrapolation debate; (5) exposures used are often at levels far in excess of those experienced by humans; (6) single-chemical exposure does not mimic human exposures, which are generally to multiple chemicals simultaneously
Mid- and short-term in vivo and in vitro bioassays	(1) more rapid and less expensive than other assays; (2) large samples that are easily replicated; (3) biologically meaningful end points are measured (mutation, *etc.*); (4) can be used as screening assays to select chemicals for long-term bioassays	(1) in vitro not fully predictive of in vivo; (2) usually organism or organ specific; (3) potencies not comparable to whole animals or humans
Chemical structure–biological activity associations	(1) relatively easy, rapid, and inexpensive; (2) reliable for certain chemical classes (*e.g.*, nitrosamines and benzidine dyes); (3) developed from biological data but not dependent on additional biological experimentation	(1) not "biological"; (2) many exceptions to formulated rules; (3) retrospective and rarely (but becoming) prospective
Mechanism-based inferences	(1) reasonably accurate for certain classes of chemicals; (2) permits refinements of hypotheses; (3) can orient risk assessments to sensitive populations	(1) mechanisms of chemical carcinogenesis undefined, multiple, and likely chemical or class specific; (2) may fail to highlight exceptions to general mechanisms

Rationale and Conceptual Basis for Genetic Toxicology Assays

Although the exact types and numbers of assays used for genetic toxicity assessment are constantly evolving and vary from country to country, the most common ones include assays for (1) gene mutation in bacteria and/or cultured mammalian cells and (2) chromosomal mutation in cultured mammalian cells and/or bone marrow within living mice. Some of the assays within this second category can also detect aneuploidy. Although these assays do not detect mutations in germ cells, they are used primarily because of the extra cost and complexity of performing germ-cell assays. Nonetheless, germ-cell assays in mice are used when information about germ-cell effects is desired.

Systematic studies over a 25-year period (1970-1995), especially at the US National Toxicology Program in North Carolina, have resulted in the use of a discrete number of assays for detecting the mutagenic activity of agents. The rationale for evaluating the usefulness of the assays was based on their ability to detect agents that cause cancer in rodents and that are suspected of causing cancer in humans (*i.e.*, carcinogens). This is because studies during the past several decades have indicated that cancer cells contain mutations in certain genes and that many carcinogens are also mutagens. Thus, cancer cells are viewed as containing somatic-cell mutations, and carcinogenesis is viewed as a type of somatic-cell mutagenesis.

The genetic toxicity assays used most commonly today have been selected not only because of their large database, relatively low cost, and ease of performance, but because they have been shown to detect many rodent and, presumptively, human carcinogens. Consequently, genetic toxicity assays are used to predict the potential carcinogenicity of agents.

An important conceptual and practical development in the field of genetic toxicology was the recognition that many carcinogens were modified by enzymes within the body, creating altered forms (metabolites) that were frequently the ultimate carcinogenic and mutagenic form of the parent chemical. To duplicate this metabolism in a petri dish, Heinrich Malling showed that the inclusion of a preparation from rodent liver contained many of the enzymes necessary to perform this metabolic conversion or activation. Thus, many genetic toxicity assays performed in dishes or tubes (in vitro) employ the addition of similar enzyme preparations. Simple preparations are called S9 mix, and purified preparations are called microsomes. Some bacterial and mammalian cells have now been genetically engineered to contain some of the genes from rodents or humans that produce these enzymes, reducing the need to add S9 mix or microsomes.

Genetic Toxicology Assays and Techniques

The primary bacterial systems used for genetic toxicity screening are the Salmonella (Ames) mutagenicity assay and, to a much lesser extent, strain WP2 of Escherichia coli. Studies in the mid-1980s indicated that the use of only two strains of the Salmonella system (TA98 and TA100) were sufficient to detect approximately 90% of the known Salmonella mutagens. Thus, these two strains are used for most screening purposes; however, various other strains are available for more extensive testing.

These assays are performed in a variety of ways, but two general procedures are the plate-incorporation and liquid-suspension assays. In the plate-incorporation assay, the cells, the test chemical and (when desired) the S9 are added together into a liquefied agar and poured onto the surface of an agar petri plate. The top agar hardens within a few minutes, and the plates are incubated for two to three days, after which time mutant cells have grown to form visually detectable clusters of cells called colonies, which are then counted. The agar medium contains selective agents or is composed of ingredients such that only the newly mutated cells will grow. The liquid-incubation assay is similar, except the cells, test agent, and S9 are incubated together in liquid that does not contain liquefied agar, and then the cells are washed free of the test agent and S9 and seeded onto the agar.

Mutations in cultured mammalian cells are detected primarily in one of two genes: hprt and tk. Similar to the bacterial assays, mammalian cell lines (developed from rodent or human cells) are exposed to the test agent in plastic culture dishes or tubes and then are seeded into culture dishes that contain medium with a selective agent that permits only mutant cells to grow. The assays used for this purpose include the CHO/HPRT, the TK6, and the mouse lymphoma L5178Y/TK$^{+/-}$ assays. Other cell lines containing various DNA repair mutations as well as containing some human genes involved in metabolism are also used. These systems permit the recovery of mutations within the gene (gene mutation) as well as mutations involving regions of the chromosome flanking the gene (chromosomal mutation). However, this latter type of mutation is recovered to a much greater extent by the tk gene systems than by the hprt gene systems due to the location of the tk gene.

Similar to the liquid-incubation assay for bacterial mutagenicity, mammalian cell mutagenicity assays generally involve the exposure of the cells in culture dishes or tubes in the presence of the test agent and S9 for several hours. The cells are then washed, cultured for several more days to allow the normal (wild-type) gene products to be degraded and the newly mutant gene products to be expressed and accumulate, and then they are seeded into medium containing a selective agent that permits only the mutant cells to grow. Like the bacterial assays, the mutant cells grow into visually detectable colonies that are then counted.

Chromosomal mutation is identified primarily by cytogenetic assays, which involve exposing rodents and/or rodent or human cells in culture dishes to a test chemical, allowing one or more cell divisions to occur, staining the chromosomes, and then visually examining the chromosomes through a microscope to detect alterations in the structure or number of chromosomes. Although a variety of endpoints can be examined, the two that are currently accepted by regulatory agencies as being the most meaningful are chromosomal aberrations and a sub-category called micronuclei.

Considerable training and expertise are required to score cells for the presence of chromosomal aberrations, making this a costly procedure in terms of time and money. In contrast, micronuclei require little training, and their detection can be automated. Micronuclei appear as small dots within the cell that are distinct from the nucleus, which contains the chromosomes. Micronuclei result from either chromosome breakage or from aneuploidy. Because of the ease of scoring micronuclei compared to chromosomal aberrations, and because recent studies indicate that agents that induce chromosomal aberrations in the bone marrow of living mice generally induce micronuclei in this tissue, micronuclei are now commonly measured as an indication of the ability of an agent to induce chromosomal mutation.

Although germ-cell assays are used far less frequently than the other assays described above, they are indispensable in determining whether an agent poses a risk to the germ cells, mutations in which can lead to health effects in succeeding generations. The most commonly used germ-cell assays are in mice, and involve systems that detect (1) heritable translocations (exchanges) among chromosomes (heritable translocation assay), (2) gene or chromosomal mutations involving specific genes (visible or biochemical specific-locus assays), and (3) mutations that affect viability (dominant lethal assay). As with the somatic-cell assays, the working assumption with the germ-cell assays is that agents positive in these assays are presumed to be potential human germ-cell mutagens.

Current Status and Future Prospects

Recent studies have indicated that only three pieces of information were necessary to detect approximately 90% of a set of 41 rodent carcinogens (*i.e.*, presumptive human carcinogens and somatic-cell mutagens). These included (1) knowledge of the chemical structure of agent, especially if it contains electrophilic moieties (see section on structure-activity relationships); (2) Salmonella mutagenicity data; and (3) data from a 90-day chronic toxicity assay in rodents (mice and rats). Indeed, essentially all of the IARC-declared human carcinogens are detectable as mutagens using just the Salmonella assay and the mouse-bone marrow micronucleus assay. The use of these mutagenicity assays for detecting potential human carcinogens is supported further by the finding that most human carcinogens are carcinogenic in both rats and mice (trans-species carcinogens) and that most trans-species carcinogens are mutagenic in Salmonella and/or induce micronuclei in mouse bone marrow.

With advances in DNA technology, the human genome project, and an improved understanding of the role of mutation in cancer, new genotoxicity assays are being developed that will likely be incorporated into standard screening procedures. Among these are the use of transgenic cells and rodents. Transgenic systems are those in which a gene from another species has been introduced into a cell or organism. For example, transgenic mice are now in experimental use that permit the detection of mutation in any organ or tissue of the animal, based on the introduction of a bacterial gene into the mouse. Bacterial cells, such as Salmonella, and mammalian cells (including human cell lines) are now available that contain genes involved in the metabolism of carcinogenic/mutagenic agents, such as the P450 genes. Molecular analysis of the actual mutations induced in the trans-gene within transgenic rodents, or within native genes such as hprt, or the target genes within Salmonella can now be performed, so that the exact nature of the mutations induced by the chemicals can be determined, providing insights into the mechanism of action of the chemical and allowing comparisons to mutations in humans presumptively exposed to the agent.

Molecular advances in cytogenetics now permit more detailed evaluation of chromosomal mutations. These include the use of probes (small pieces of DNA) that attach (hybridize) to specific genes. Rearrangements of genes on the chromosome can then be revealed by the altered location of the probes, which are fluorescent and easily visualized as colored sectors on the chromosomes. The single-cell gel electrophoresis assay for DNA breakage (commonly called the "comet" assay) permits the detection of DNA breaks within single cells and may become an extremely useful tool in combination with cytogenetic techniques for detecting chromosomal damage.

After many years of use and the generation of a large and systematically developed database, genetic toxicity assessment can now be done with just a few assays for relatively small cost in a short period of time (a few weeks). The data produced can be used to predict the ability of an agent to be a rodent and, presumptively, human carcinogen/somatic-cell mutagen. Such an ability makes it possible to limit the introduction into the environment of mutagenic and carcinogenic agents and to develop alternative, nonmutagenic agents. Future studies should lead to even better methods with greater predictivity than the current assays.

IN VITRO TOXICITY TESTING

The emergence of sophisticated technologies in molecular and cellular biology has spurred a relatively rapid evolution in the life sciences, including toxicology. In effect, the focus of toxicology is shifting from whole animals and populations of whole animals to the cells and molecules of individual animals and humans. Since the mid-1980s, toxicologists have begun to employ these new methodologies in assessing the effects of chemicals on living systems. As a logical progression, such methods are being adapted for the purposes of toxicity testing. These scientific advances have worked together with social and economic factors to effect change in the evaluation of product safety and potential risk.

Economic factors are specifically related to the volume of materials that must be tested. A plethora of new cosmetics, pharmaceuticals, pesticides, chemicals and household products is introduced into the market every year. All of these products must be evaluated for their potential toxicity. In addition, there is a backlog of chemicals already in use that have not been adequately tested. The enormous task of obtaining detailed safety information on all of these chemicals using traditional whole animal testing methods would be costly in terms of both money and time, if it could even be accomplished.

There are also societal issues that relate to public health and safety, as well as increasing public concern about the use of animals for product safety testing. With regard to human safety, public interest and environmental advocacy groups have placed significant pressure on government agencies to apply more stringent regulations on chemicals. A recent example of this has been a movement by some environmental groups to ban chlorine and chlorine-containing compounds in the United States. One of the motivations for such an extreme action lies in the fact that most of these compounds have never been adequately tested. From a toxicological perspective, the concept of banning a whole class of diverse chemicals based simply on the presence of chlorine is both scientifically unsound and irresponsible. Yet, it is understandable that from the public's perspective, there must be some assurance that chemicals released into the environment do not pose a significant health risk. Such a situation underscores the need for more efficient and rapid methods to assess toxicity.

The other societal concern that has impacted the area of toxicity testing is animal welfare. The growing number of animal protection groups throughout the world have voiced considerable opposition to the use of whole animals for product safety testing. Active campaigns have been waged against manufacturers of cosmetics, household and personal care products and pharmaceuticals in attempts to stop animal testing. Such efforts in Europe have resulted in the passage of the Sixth Amendment to Directive 76/768/EEC (the Cosmetics Directive). The consequence of this Directive is that cosmetic products or cosmetic ingredients that have been tested in animals after January 1, 1998 cannot be marketed in the European Union, unless alternative methods are insufficiently validated. While this Directive has no jurisdiction over the sale of such products in the United States or other countries, it will significantly affect those companies that have international markets that include Europe.

The concept of alternatives, which forms the basis for the development of tests other than those on whole animals, is defined by the three Rs: reduction in the numbers of animals used; refinement of protocols so that animals experience less stress or discomfort; and replacement of current animal tests with in vitro tests (*i.e.*, tests done outside of the living animal), computer models or test on lower vertebrate or invertebrate species. The three Rs were introduced in a book published in 1959 by two British scientists, W.M.S. Russell and Rex Burch, The Principles of Humane Experimental Technique. Russell and Burch maintained that the only way in which valid scientific results could be obtained is through the humane treatment of animals, and believed that methods should be developed to reduce animal use and ultimately replace it. Interestingly, the principles outlined by Russell and Burch received little attention until the resurgence of the animal welfare movement in the mid-1970s. Today the concept of the three Rs is very much in the forefront with regard to research, testing and education.

In summary, the development of in vitro test methodologies has been influenced by a variety of factors that have converged over the last ten to 20 years. It is difficult to ascertain if any of these factors alone would have had such a profound effect on toxicity testing strategies.

Concept of in Vitro Toxicity Tests

This section will focus solely on in vitro methods for evaluating toxicity, as one of the alternatives to whole-animal testing. Additional non-animal alternatives such as computer modelling and quantitative structure-activity relationships are discussed in other articles of this chapter.

In vitro studies are generally conducted in animal or human cells or tissues outside of the body. In vitro literally means "in glass", and refers to procedures carried out on living material or components of living material cultured in petri dishes or in test tubes under defined conditions. These may be contrasted with in vivo studies, or those carried out "in the living animal". While it is difficult, if not impossible, to project the effects of a chemical on a complex organism when the observations are confined to a single type of cells in a dish, in vitro studies do provide a significant amount of information about intrinsic toxicity as well as cellular and molecular mechanisms of toxicity. In addition, they offer many advantages over in vivo studies in that they are generally less expensive and they may be conducted under more controlled conditions. Furthermore, despite the fact that small numbers of animals are still needed to obtain cells for in vitro cultures, these methods may be considered reduction alternatives (since many fewer animals are used compared to in vivo studies) and refinement alternatives (because they eliminate the need to subject the animals to the adverse toxic consequences imposed by in vivo experiments).

In order to interpret the results of in vitro toxicity tests, determine their potential usefulness in assessing toxicity and relate them to the overall toxicological process in vivo, it is necessary to understand which part of the toxicological process is being examined. The entire toxicological process consists of events that begin with the organism's exposure to a physical or chemical agent, progress through cellular and molecular interactions and ultimately manifest themselves in the response of the whole organism. In vitro tests are generally limited to the part of the toxicological process that takes place at the cellular and molecular level. The types of information that may be obtained from in vitro studies include pathways of metabolism, interaction of active metabolites with cellular and molecular targets and potentially measurable toxic endpoints that can serve as molecular biomarkers for exposure. In an ideal situation, the mechanism of toxicity of each chemical from exposure to organismal manifestation would be known, such that the information obtained from in vitro tests could be fully interpreted and related to the response of the whole organism. However, this is virtually impossible, since relatively few complete toxicological mechanisms have been elucidated. Thus, toxicologists are faced with a situation in which the results of an in vitro test cannot be used as an entirely accurate prediction of in vivo toxicity because the mechanism is unknown. However, frequently during the process of developing an in vitro test, components of the cellular and molecular mechanism(s) of toxicity are elucidated.

One of the key unresolved issues surrounding the development and implementation of in vitro tests is related to the following consideration: should they be mechanistically based or is it sufficient for them to be descriptive? It is inarguably better from a scientific perspective to utilize only mechanistically based tests as replacements for in vivo tests. However in the absence of complete mechanistic knowledge, the prospect of developing in vitro tests to completely replace whole animal tests in the near future is almost nil. This does not, however, rule out the use of more descriptive types of assays as early screening tools, which is the case presently. These screens have resulted in a significant reduction in animal use. Therefore, until such time as more mechanistic information is generated, it may be necessary to employ to a more limited extent, tests whose results simply correlate well with those obtained in vivo.

In Vitro Tests for Cytotoxicity

In this section, several in vitro tests that have been developed to assess a chemical's cytotoxic potential will be described. For the most part, these tests are easy to perform and analysis can be automated. One commonly used in vitro test for cytotoxicity is the neutral red assay. This assay is done on cells in culture, and for most applications, the cells can be maintained in culture dishes that contain 96 small wells, each 6.4 mm in diameter. Since each well can be used for a single determination, this arrangement can accommodate multiple concentrations of the test chemical as well as positive and negative controls with a sufficient number of replicates for each. Following treatment of the cells with various concentrations of the test chemical ranging over at least two orders of magnitude (*e.g.*, from 0.01 mM to 1 mM), as well as positive and negative control chemicals, the cells are rinsed and treated with neutral red, a dye that can be taken up and retained only by live cells. The dye may be added upon removal of the test chemical to determine immediate effects, or it may be added at various times after the test chemical is removed to determine cumulative or delayed effects. The intensity of the colour in each well corresponds to the number of live cells in that well. The colour intensity is measured by a spectrophotometer which may be equipped with a plate reader. The plate reader is programmed to provide individual measure-ments for each of the 96 wells of the culture dish. This automated methodology permits the investigator to rapidly perform a concentration-response experiment and to obtain statistically useful data.

Another relatively simple assay for cytotoxicity is the MTT test. MTT (3[4,5-dimethylthiazol-2-yl]-2,5-diphenyltetrazolium bromide) is a tetrazolium dye that is reduced by mitochondrial enzymes to a blue colour. Only cells with viable mitochondria will retain the ability to carry out this reaction; therefore the colour intensity is directly related to the degree of mitochondrial integrity. This is a useful test to detect general cytotoxic compounds as well as those agents that specifically target mitochondria.

The measurement of lactate dehydrogenase (LDH) activity is also used as a broad-based assay for cytotoxicity. This enzyme is normally present in the cytoplasm of living cells and is released into the cell culture medium through leaky cell membranes of dead or dying cells that have been adversely affected by a toxic agent. Small amounts of culture medium may be removed at various times after chemical treatment of the cells to measure the amount of LDH released and determine a time course of toxicity. While the LDH release assay is a very general assessment of cytotoxicity, it is useful because it is easy to perform and it may be done in real time.

There are many new methods being developed to detect cellular damage. More sophisticated methods employ fluorescent probes to measure a variety of intracellular parameters, such as calcium release and changes in pH and mem-brane potential. In general, these probes are very sensitive and may detect more subtle cellular changes, thus reducing the need to use cell death as an endpoint. In addition, many of these fluorescent assays may be automated by the use of 96-well plates and fluorescent plate readers.

Once data have been collected on a series of chemicals using one of these tests, the relative toxicities may be determined. The relative toxicity of a chemical, as determined in an in vitro test, may be expressed as the concentration that exerts a 50% effect on the endpoint response of untreated cells. This determination is referred to as the EC_{50} (Effective Concentration for 50% of the cells) and may be used to compare toxicities of different chemicals in vitro. (A similar term used in evaluating relative toxicity is IC_{50}, indicating the concentration of a chemical that causes a 50% inhibition of a cellular process, *e.g.*, the ability to take up neutral red.) It is not easy to assess whether the relative in vitro toxicity of the chemicals is comparable to their relative in vivo toxicities, since there are so many confounding factors in the in vivo system, such as toxicokinetics, metabolism, repair and defence mechanisms. In addition, since most of these assays measure general cytotoxicity endpoints, they are not mechanistically based. Therefore, agreement between in vitro and in vivo relative toxicities is simply correlative. Despite the numerous complexities and difficulties in extrapolating from in vitro to in vivo, these in vitro tests are proving to be very valuable because they are simple and inexpensive to perform and may be used as screens to flag highly toxic drugs or chemicals at early stages of development.

Target Organ Toxicity

In vitro tests can also be used to assess specific target organ toxicity. There are a number of difficulties associated with designing such tests, the most notable being the inability of in vitro systems to maintain many of the features of the organ in vivo. Frequently, when cells are taken from animals and placed into culture, they tend either to degenerate quickly and/or to dedifferentiate, that is, lose their organ-like functions and become more generic. This presents a problem in that within a short period of time, usually a few days, the cultures are no longer useful for assessing organ-specific effects of a toxin.

Many of these problems are being overcome because of recent advances in molecular and cellular biology. Information that is obtained about the cellular environment in vivo may be utilized in modulating culture conditions in vitro. Since the mid-1980s, new growth factors and cytokines have been discovered, and many of these are now available commercially. Addition of these factors to cells in culture helps to preserve their integrity and may also help to retain more differentiated functions for longer periods of time. Other basic studies have increased the knowledge of the nutritional and hormonal requirements of cells in culture, so that new media may be formulated. Recent advances have also been made in identifying both naturally occurring and artificial extracellular matrices on which cells may be cultured. Culture of cells on these different matrices can have profound effects on both their structure and function. A major advantage derived from this knowledge is the ability to intricately control the environment of cells in culture and individually examine the effects of these factors on basic cell processes and on their responses to different chemical agents. In short, these systems can provide great insight into organ-specific mechanisms of toxicity.

Many target organ toxicity studies are conducted in primary cells, which by definition are freshly isolated from an organ, and usually exhibit a finite lifetime in culture. There are many advantages to having primary cultures of a single cell type from an organ for toxicity assessment. From a mechanistic perspective, such cultures are useful for studying specific cellular targets of a chemical. In some instances, two or more cell types from an organ may be cultured together, and this provides an added advantage of being able to look at cell-cell interactions in response to a toxin. Some co-culture systems for skin have been engineered so that they form a three dimensional structure resembling skin in vivo. It is also possible to co-culture cells from different organs — for example, liver and kidney. This type of culture would be useful in assessing the effects specific to kidney cells, of a chemical that must be bioactivated in the liver.

Molecular biological tools have also played an important role in the development of continuous cell lines that can be useful for target organ toxicity testing. These cell lines are generated by transfecting DNA into primary cells. In the transfection procedure, the cells and the DNA are treated such that the DNA can be taken up by the cells. The DNA is usually from a virus and contains a gene or genes that, when expressed, allow the cells to become immortalized (*i.e.*, able to live and grow for extended periods of time in culture). The DNA can also be engineered so that the immortalizing gene is controlled by an inducible promoter. The advantage of this type of construct is that the cells will divide only when they receive the appropriate chemical stimulus to allow expression of the immortalizing gene. An example of such a construct is the large T antigen gene from Simian Virus 40 (SV40) (the immortalizing gene), preceded by the promoter region of the metallothionein gene, which is induced by the presence of a metal in the culture medium. Thus, after the gene is transfected into the cells, the cells may be treated with low concentrations of zinc to stimulate the MT promoter and turn on the expression of the T antigen gene. Under these conditions, the cells proliferate. When zinc is removed from the medium, the cells stop dividing and under ideal conditions return to a state where they express their tissue-specific functions.

The ability to generate immortalized cells combined with the advances in cell culture technology have greatly contributed to the creation of cell lines from many different organs, including brain, kidney and liver. However, before these cell lines may be used as a surrogate for the bona fide cell types, they must be carefully characterized to determine how "normal" they really are.

Other in vitro systems for studying target organ toxicity involve increasing complexity. As in vitro systems progress in complexity from single cell to whole organ culture, they become more comparable to the in vivo milieu, but at the same time they become much more difficult to control given the increased number of variables. Therefore, what may be gained in moving to a higher level of organization can be lost in the inability of the researcher to control the experimental environment. Table compares some of the characteristics of various in vitro systems that have been used to study hepatotoxicity.

Table. Comparison of in vitro systems for hepatotoxicity studies

System	Complexity (level of interaction)	Ability to retain liver-specific functions	Potential duration of culture	Ability to control environment
Immortalized cell lines	some cell to cell (varies with cell line)	poor to good (varies with cell line)	indefinite	excellent
Primary hepatocyte cultures	cell to cell	fair to excellent (varies with culture conditions)	days to weeks	excellent
Liver cell co-cultures	cell to cell (between the same and different cell types)	good to excellent	weeks	excellent
Liver slices	cell to cell (among all cell types)	good to excellent	hours to days	good
Isolated, perfused liver	cell to cell (among all cell types), and intra-organ	excellent	hours	fair

Precision-cut tissue slices are being used more extensively for toxicological studies. There are new instruments available that enable the researcher to cut uniform tissue slices in a sterile environment. Tissue slices offer some advantage over cell culture systems in that all of the cell types of the organ are present and they maintain their in vivo architecture and intercellular communication. Thus, in vitro studies may be conducted to determine the target cell type within an organ as well as to investigate specific target organ toxicity. A disadvantage of the slices is that they degenerate rapidly after the first 24 hours of culture, mainly due to poor diffusion of oxygen to the cells on the interior of the slices. However, recent studies have indicated that more efficient aeration may be achieved by gentle rotation. This, together with the use of a more complex medium, allows the slices to survive for up to 96 hours.

Tissue explants are similar in concept to tissue slices and may also be used to determine the toxicity of chemicals in specific target organs. Tissue explants are established by removing a small piece of tissue (for teratogenicity studies, an intact embryo) and placing it into culture for further study. Explant cultures have been useful for short-term toxicity studies including irritation and corrosivity in skin, asbestos studies in trachea and neurotoxicity studies in brain tissue.

Isolated perfused organs may also be used to assess target organ toxicity. These systems offer an advantage similar to that of tissue slices and explants in that all cell types are present, but without the stress to the tissue introduced by the manipulations involved in preparing slices. In addition, they allow for the maintenance of intra-organ interactions. A major disadvantage is their short-term viability, which limits their use for in vitro toxicity testing. In terms of serving as an alternative, these cultures may be considered a refinement since the animals do not experience the adverse consequences of in vivo treatment with toxicants. However, their use does not significantly decrease the numbers of animals required.

In summary, there are several types of in vitro systems available for assessing target organ toxicity. It is possible to acquire much information about mechanisms of toxicity using one or more of these techniques. The difficulty remains in knowing how to extrapolate from an in vitro system, which represents a relatively small part of the toxicological process, to the whole process occurring in vivo.

In Vitro Tests for Ocular Irritation

Perhaps the most contentious whole-animal toxicity test from an animal welfare perspective is the Draize test for eye irritation, which is conducted in rabbits. In this test, a small fixed dose of a chemical is placed in one of the rabbit's eyes while the other eye is used as a control. The degree of irritation and inflammation is scored at various times after exposure. A major effort is being made to develop methodologies to replace this test, which has been criticized not only for humane reasons, but also because of the subjectivity of the observations and variability of the results. It is interesting to note that despite the harsh criticism the Draize test has received, it has proven to be remarkably successful in predicting human eye irritants, particularly slightly to moderately irritating substances, that are difficult to identify by other methods. Thus, the demands on in vitro alternatives are great.

The quest for alternatives to the Draize test is a complicated one, albeit one that is predicted to be successful. Numerous in vitro and other alternatives have been developed and in some cases they have been implemented. Refinement alternatives to the Draize test, which by definition, are less painful or distressful to the animals, include the Low Volume Eye Test, in which smaller amounts of test materials are placed in the rabbits' eyes, not only for humane reasons, but to more closely mimic the amounts to which people may actually be accidentally exposed. Another refinement is that substances which have a pH less than 2 or greater than 11.5 are no longer tested in animals since they are known to be severely irritating to the eye.

Between 1980 and 1989, there has been an estimated 87% decline in the number of rabbits used for eye irritation testing of cosmetics. In vitro tests have been incorporated as part of a tier-testing approach to bring about this vast reduction in whole-animal tests. This approach is a multi-step process that begins with a thorough examination of the historical eye irritation data and physical and chemical analysis of the chemical to be evaluated. If these two processes do not yield enough information, then a battery of in vitro tests is performed. The additional data obtained from the in vitro tests might then be sufficient to assess the safety of the substance. If not, then the final step would be to perform limited in vivo tests. It is easy to see how this approach can eliminate or at least drastically reduce the numbers of animals needed to predict the safety of a test substance.

The battery of in vitro tests that is used as part of this tier-testing strategy depends upon the needs of the particular industry. Eye irritation testing is done by a wide variety of industries from cosmetics to pharmaceuticals to industrial chemicals. The type of information required by each industry varies and therefore it is not possible to define a single battery of in vitro tests. A test battery is generally designed to assess five parameters: cytotoxicity, changes in tissue physiology and biochemistry, quantitative structure-activity relationships, inflammation mediators, and recovery and repair. An example of a test for cytotoxicity, which is one possible cause for irritation, is the neutral red assay using cultured cells (see above). Changes in cellular physiology and biochemistry resulting from exposure to a chemical may be assayed in cultures of human corneal epithelial cells. Alternatively, investigators have also used intact or dissected bovine or chicken eyeballs obtained from slaughterhouses. Many of the endpoints measured in these whole organ cultures are the same as those measured in vivo, such as corneal opacity and corneal swelling.

Inflammation is frequently a component of chemical-induced eye injury, and there are a number of assays available to examine this parameter. Various biochemical assays detect the presence of mediators released during the inflammatory process such as arachidonic acid and cytokines. The chorioallantoic membrane (CAM) of the hen's egg may also be used as an indicator of inflammation. In the CAM assay, a small piece of the shell of a ten-to-14-day chick embryo is removed to expose the CAM. The chemical is then applied to the CAM and signs of inflammation, such as vascular hemorrhaging, are scored at various times thereafter.

One of the most difficult in vivo processes to assess in vitro is recovery and repair of ocular injury. A newly developed instrument, the silicon microphysiometer, measures small changes in extracellular pH and can been used to monitor cultured cells in real time. This analysis has been shown to correlate fairly well with in vivo recovery and has been used as an in vitro test for this process. This has been a brief overview of the types of tests being employed as alternatives to the Draize test for ocular irritation. It is likely that within the next several years a complete series of in vitro test batteries will be defined and each will be validated for its specific purpose.

Validation

The key to regulatory acceptance and implementation of in vitro test methodologies is validation, the process by which the credibility of a candidate test is established for a specific purpose. Efforts to define and coordinate the validation process have been made both in the United States and in Europe. The European Union established the European Centre for the Validation of Alternative Methods (ECVAM) in 1993 to coordinate efforts there and to interact with American organizations such as the Johns Hopkins Centre for Alternatives to Animal Testing (CAAT), an academic centre in the United States, and the Interagency Coordinating Committee for the Validation of Alternative Methods (ICCVAM), composed of representatives from the National Institutes of Health, the US Environmental Protection Agency, the US Food and Drug Administration and the Consumer Products Safety Commission.

Validation of in vitro tests requires substantial organization and planning. There must be consensus among government regulators and industrial and academic scientists on acceptable procedures, and sufficient oversight by a scientific advisory board to ensure that the protocols meet set standards. The validation studies should be performed in a series of reference laboratories using calibrated sets of chemicals from a chemical bank and cells or tissues from a single source. Both intralaboratory repeatability and interlaboratory reproducibility of a candidate test must be demonstrated and the results subjected to appropriate statistical analysis. Once the results from the different components of the validation studies have been compiled, the scientific advisory board can make recommendations on the validity of the candidate test(s) for a specific purpose. In addition, results of the studies should be published in peer-reviewed journals and placed in a database.

The definition of the validation process is currently a work in progress. Each new validation study will provide information useful to the design of the next study. International communication and cooperation are essential for the expeditious development of a widely acceptable series of protocols, particularly given the increased urgency imposed by the passage of the EC Cosmetics Directive. This legislation may indeed provide the needed impetus for a serious validation effort to be undertaken. It is only through completion of this process that the acceptance of in vitro methods by the various regulatory communities can commence.

Conclusion

This article has provided a broad overview of the current status of in vitro toxicity testing. The science of in vitro toxicology is relatively young, but it is growing exponentially. The challenge for the years ahead is to incorporate the mechanistic knowledge generated by cellular and molecular studies into the vast inventory of in vivo data to provide a more complete description of toxicological mechanisms as well as to establish a paradigm by which in vitro data may be used to predict toxicity in vivo. It will only be through the concerted efforts of toxicologists and government representatives that the inherent value of these in vitro methods can be realized.

STRUCTURE ACTIVITY RELATIONSHIPS

Structure activity relationships (SAR) analysis is the utilization of information on the molecular structure of chemicals to predict important characteristics related to persistence, distribution, uptake and absorption, and toxicity. SAR is an alternative method of identifying potential hazardous chemicals, which holds promise of assisting industries and governments in prioritizing substances for further evaluation or for early-stage decision making for new chemicals. Toxicology is an increasingly expensive and resource-intensive undertaking. Increased concerns over the potential for chemicals to cause adverse effects in exposed human populations have prompted regulatory and health agencies to expand the range and sensitivity of tests to detect toxicological hazards. At the same time, the real and perceived burdens of regulation upon industry have provoked concerns for the practicality of toxicity testing methods and data analysis. At present, the determination of chemical carcinogenicity depends upon lifetime testing of at least two species, both sexes, at several doses, with careful histopathological analysis of multiple organs, as well as detection of preneoplastic changes in cells and target organs. In the United States, the cancer bioassay is estimated to cost in excess of $3 million (1995 dollars).

Even with unlimited financial resources, the burden of testing the approximately 70,000 existing chemicals produced in the world today would exceed the available resources of trained toxicologists. Centuries would be required to complete even a first tier evaluation of these chemicals (NRC 1984). In many countries ethical concerns over the use of animals in toxicity testing have increased, bringing additional pressures upon the uses of standard methods of toxicity testing. SAR has been widely used in the pharmaceutical industry to identify molecules with potential for beneficial use in treatment (Hansch and Zhang 1993). In environmental and occupational health policy, SAR is used to predict the dispersion of compounds in the physical-chemical environment and to screen new chemicals for further evaluation of potential toxicity. Under the US Toxic Substances Control Act (TSCA), the EPA has used since 1979 an SAR approach as a "first screen" of new chemicals in the premanufacture notification (PMN) process; Australia uses a similar approach as part of its new chemicals notification (NICNAS) procedure. In the US SAR analysis is an important basis for determining that there is a reasonable basis to conclude that manufacture, processing, distribution, use or disposal of the substance will present an unreasonable risk of injury to human health or the environment, as required by Section 5(f) of TSCA. On the basis of this finding, EPA can then require actual tests of the substance under Section 6 of TSCA.

Rationale for SAR

The scientific rationale for SAR is based upon the assumption that the molecular structure of a chemical will predict important aspects of its behaviour in physical-chemical and biological systems.

SAR Process

The SAR review process includes identification of the chemical structure, including empirical formulations as well as the pure compound; identification of structurally analogous substances; searching databases and literature for information on structural analogs; and analysis of toxicity and other data on structural analogs. In some rare cases, information on the structure of the compound alone can be sufficient to support some SAR analysis, based upon well-understood mechanisms of toxicity. Several databases on SAR have been compiled, as well as computer-based methods for molecular structure prediction.

With this information, the following endpoints can be estimated with SAR:

- physical-chemical parameters: boiling point, vapour pressure, water solubility, octanol/water partition coefficient

- biological/environmental fate parameters: biodegradation, soil sorption, photodegradation, pharmacokinetics
- toxicity parameters: aquatic organism toxicity, absorption, acute mammalian toxicity (limit test or LD50), dermal, lung and eye irritation, sensitization, subchronic toxicity, mutagenicity.

It should be noted that SAR methods do not exist for such important health endpoints as carcinogenicity, developmental toxicity, reproductive toxicity, neurotoxicity, immunotoxicity or other target organ effects. This is due to three factors: the lack of a large database upon which to test SAR hypotheses, lack of knowledge of structural determinants of toxic action, and the multiplicity of target cells and mechanisms that are involved in these endpoints (see "The United States approach to risk assessment of reproductive toxicants and neurotoxic agents"). Some limited attempts to utilize SAR for predicting pharmacokinetics using information on partition coefficients and solubility. More extensive quantitative SAR has been done to predict P450-dependent metabolism of a range of compounds and binding of dioxin- and PCB-like molecules to the cytosolic "dioxin" receptor.

SAR has been shown to have varying predictability for some of the endpoints listed above, as shown in table. This table presents data from two comparisons of predicted activity with actual results obtained by empirical measurement or toxicity testing. SAR as conducted by US EPA experts performed more poorly for predicting physical-chemical properties than for predicting biological activity, including biodegradation. For toxicity endpoints, SAR performed best for predicting mutagenicity. Ashby and Tennant (1991) in a more extended study also found good predictability of short-term genotoxicity in their analysis of NTP chemicals. These findings are not surprising, given current understanding of molecular mechanisms of genotoxicity (see "Genetic toxicology") and the role of electrophilicity in DNA binding. In contrast, SAR tended to underpredict systemic and subchronic toxicity in mammals and to overpredict acute toxicity to aquatic organisms.

Table Comparison of SAR and test data: OECD/NTP analyses

Endpoint	Agreement (%)	Disagreement (%)	Number
Boiling point	50	50	30
Vapour pressure	63	37	113
Water solubility	68	32	133
Partition coefficient	61	39	82
Biodegradation	93	7	107
Fish toxicity	77	22	130
Daphnia toxicity	67	33	127
Acute mammalian toxicity (LD_{50})	80	20[1]	142
Skin irritation	82	18	144

Eye irritation	78	22	144
Skin sensitization	84	16	144
Subchronic toxicity	57	32	143
Mutagenicity[2]	88	12	139
Mutagenicity[3]	82–94[4]	1–10	301
Carcinogenicity[3] : Two year bio-assay	72–95[4]	–	301

Source: Data from OECD, personal communication C. Auer,US EPA. Only those endpoints for which comparable SAR predictions and actual test data were available were used in this analysis. NTP data are from Ashby and Tennant 1991.

- Of concern was the failure by SAR to predict acute toxicity in 12% of the chemicals tested.

- OECD data, based on Ames test concordance with SAR

- NTP data, based on genetox assays compared to SAR predictions for several classes of "structurally alerting chemicals".

- Concordance varies with class; highest concordance was with aromatic amino/nitro compounds; lowest with "miscellaneous" structures.

For other toxic endpoints, as noted above, SAR has less demonstrable utility. Mammalian toxicity predictions are complicated by the lack of SAR for toxicokinetics of complex molecules. Nevertheless, some attempts have been made to propose SAR principles for complex mammalian toxicity endpoints (for instance, see Bernstein (1984) for an SAR analysis of potential male reproductive toxicants). In most cases, the database is too small to permit rigorous testing of structure-based predictions.

At this point it may be concluded that SAR may be useful mainly for prioritizing the investment of toxicity testing resources or for raising early concerns about potential hazard. Only in the case of mutagenicity is it likely that SAR analysis by itself can be utilized with reliability to inform other decisions. For no endpoint is it likely that SAR can provide the type of quantitative information required for risk assessment purposes as discussed elsewhere in this chapter and Encyclopaedia.

TOXICOLOGY IN HEALTH AND SAFETY REGULATION

Toxicology plays a major role in the development of regulations and other occupational health policies. In order to prevent occupational injury and illness, decisions are increasingly based upon information obtainable prior to or in the absence of the types of human exposures that would yield definitive information on risk such as epidemiology studies. In addition, toxicological studies, as described in this chapter, can provide precise information on dose and response under the controlled conditions of laboratory research; this information is often difficult to obtain in the uncontrolled setting of occupational exposures. However, this information must be carefully evaluated in order to estimate the likelihood of adverse effects in humans, the nature of these adverse effects, and the quantitative relationship between exposures and effects.

Considerable attention has been given in many countries, since the 1980s, to developing objective methods for utilizing toxicological information in regulatory decision-making. Formal methods, frequently referred to as risk assessment, have been proposed and utilized in these countries by both governmental and non-governmental entities. Risk assessment has been varyingly defined; fundamentally it is an evaluative process that incorporates toxicology, epidemiology and exposure information to identify and estimate the probability of adverse effects associated with exposures to hazardous substances or conditions. Risk assessment may be qualitative in nature, indicating the nature of an adverse effect and a general estimate of likelihood, or it may be quantitative, with estimates of numbers of affected persons at specific levels of exposure. In many regulatory systems, risk assessment is undertaken in four stages: hazard identification, the description of the nature of the toxic effect; dose-response evaluation, a semi-quantitative or quantitative analysis of the relationship between exposure (or dose) and severity or likelihood of toxic effect; exposure assessment, the evaluation of information on the range of exposures likely to occur for populations in general or for sub-groups within populations; risk characterization, the compilation of all the above information into an expression of the magnitude of risk expected to occur under specified exposure conditions (see NRC 1983 for a statement of these principles).

In this section, three approaches to risk assessment are presented as illustrative.It is impossible to provide a comprehensive compendium of risk assessment methods used throughout the world, and these selections should not be taken as prescriptive. It should be noted that there are trends towards harmonization of risk assessment methods, partly in response to provisions in the recent GATT accords. Two processes of international harmonization of risk assessment methods are currently underway, through the International Programme on Chemical Safety (IPCS) and the Organization for Economic Cooperation and Development (OECD). These organizations also maintain current information on national approaches to risk assessment.

How Toxicants Enter Biological Organisms

Forhigherorderorganismsthepathofthechemicalagentthroughthebodyiswelldefined. After the toxicant enters the organism, it moves into the bloodstream and is eventually eliminated or it is transported to the target organ. The damage is exerted at the target organ.

A common misconception is that damage occurs in the organ where the toxicant is most concentrated. Lead, for instance, is stored in humans mostly in the bone structure, but the damage occurs in many organs. For corrosive chemicals the damage to the organism can occur without absorption or transport through the bloodstream.

Toxicants enter biological organisms by the following routes:

- Ingestion: through the mouth into the stomach
- Inhalation: through the mouth or nose into the lungs
- Injection: through cuts into the skin
- Dermal absorption: through skin membrane

All of the above entry routes are controlled by the application of proper industrial hygiene techniques. These control techniques will be discussed in more detail in industrial hygiene. Of the four routes of entry, the inhalation and dermal routes are the most significant to industrial facilities. Inhalation is the easiest to quantify by the direct measurement of airborne concentrations; the usual exposure is by vapour, but small solid and liquid particles can also contribute.

Injection, inhalation and dermal absorption generally result in the toxicant entering the blood stream unaltered. Toxicants entering through ingestion are frequently modified or excreted.

Outline the Need for Excretion in All Living Organisms

- The idea behind excretion holds true for all living organisms. If excess toxins, salts, and water builds up it may cause serious bodily harm to that organism. The kidneys make sure that this doesn't happen and allow for normal bodily functions to continue. If not for this system the organism may also end up needing to spend excessive amounts of energy to filter out the toxins that could kill it.

State that excretory products in plants include oxygen and carbon dioxide, and in animals, they include carbon dioxide from respiration, and nitrogenous compounds.

Excretory products in plants include oxygen, and in animals they include nitrogenous compounds and carbon dioxide.

Discuss the relationship between the different nitrogenous waste products and habitat in mammals, birds, and reptiles

- Any metabolism involving proteins and nucleic acids produces a nitrogen containing waste with the main waste product being Ammonia. The direct excretion of Ammonia, in one form or another, is the most biologically efficient way of disposing of waste. In fish ammonia can simply be excreted purely as it is extremely water soluble. Fishes mainly lose Ammonia through the epithelium of their gills as $NH4+$. Fresh water fishes also take up sodium as an exchange for the Ammonia. Animals, however, can not excrete pure Ammonia since it is too toxic to be excreted without first being diluted and terrestrial animals can't dispose of it quickly enough. Thus mammals excrete urea, a product which is much less toxic than Ammonia and requires less water to be excreted which is important to terrestrial creatures where water is more scarce. Finally birds excrete Uric Acid as their form of Ammonia removal. It is much less soluble in water than Ammonia is, thus leaving much more water for the bird which is needed during long migration patterns. Uric acid can also precipitate, allowing for shelled offspring who are permeable to gas but not permeable to liquid, as Ammonia or Urea would be.

The Human Kidney

Define Excretion

- Excretion is the removal of waste products of metabolic pathways from the body

Draw (and label) the structure of the kidney

- Things to include - Cortex, Medula, Ureter, Renal Blood Vessels, and Pelvis

Draw the structure of the glomerulus and associated nephron to show the function of each part

Explain the process of ultra-filtration

- Blood pressure forces fluid from the capillaries of the Glomerulus across the Epithelium of the Bowman's Capsule, and into the lumen of Nephron tubule. The porous capillaries act as a filter as they have membranes which are permeable to water and small solutes but not to blood cells or larger molecules. The efferent arteriole reabsorbs amino acids, glucose, water, and salts and NH2, excess water, and sodium chloride descend into the loop of Henle. There more water and sodium is absorbed and tubular excretion begins in distal convoluted tube. Large molecules which weren't absorbed enter for excretion and the collecting duct takes it to the renal pyramid.

- 15-20% of blood plasma is filtered into glomerulus, that exit through fenestrated blood capillaries (porous). The basement membrane is a protein layer between the glomerulus and Bowman's Capsule that prevents blood cells and large proteins from entering the Bowman's Capsule.

Define osmoregulation

- Osmoregulation - the control of water and solute levels

Explain the reabsorption of glucose, water, and salts in the proximal convoluted tube

- In the Proximal Convoluted Tubule there is selective re-absorption of water and salt, depending on how dehydrated the body is. In the presence of ADH, much more water is reabsorbed than when ADH isn't present. Microvilli also helps to expand surface area to allow for more absorption. Filtration, on the other hand, is not selective at all as it is important that essential nutrients return to the body so sugar, vitamins, and other organic nutrients are reabsorbed into the filtrate. Some are actively transported (glucose, amino acids, sodium ions) and some are passively transported (H_2O, potassium ions, chloride ions).

Explain the roles of the Loop of Henle, medulla, collecting duct and ADH in maintaining the water balance of the blood

- Water leaves the Nephron *via* Osmosis while in the descending limb of the Henle's loop due to the increasing salt concentration of the medulla in the kidney. Blood will pass into the capillaries where it is removed and salt will diffuse into the filtrate. The ascending limb is impermeable to water, allowing various salts to leave. Fluid that leaves the loop is less concentrated than the tissue around it, and blood entering the medulla will lose water due to osmosis and pick up salt and urea through diffusion. The exact opposite happens in the ascending capillary. When a lack of water is detected in the kidney, ADH is released by the pituitary gland which increases permeability of the walls in the distal convoluted tube and the collecting duct. When ADH isn't present, walls are impermeable and water is used to dilute urine.

Compare the composition of blood in the renal artery and renal vein, and compare the composition of the glomerular filtrate and urine

- Renal Artery – Large molecules, More Toxins, Oxygenated, More Salt/ Ions, More H2O, Less CO2, More Nutrients, Supplies Kidneys with O2

- Renal Vein - Less Toxins, Deoxygenated, Less Salt, Less H2O, More CO2, Less Nutrients, Returns blood from Kidneys to heart, blood contains no wastes, with less O2, urea, salt/ions, and more CO2

- Glomerular Filtrate: Small molecules, Amino acids, glucose, Na+ Cl, NH2, Urea, Nutrient excess, Nitrogenous waste, H+, H20, hormones, vitamins

- Urine: Wastes: Urea, Na+, Cl, NH2, H2O, large molecules

Outline the structure and action of kidney dialysis machines

- Blood enters machine from patient's vein and runs through partially permeable tubules and into the dialysis chamber. The tube allows nitrogenous waste to diffuse from the blood and into the dialysis fluid. Urea diffuses through the membrane, as dialysis fluid contains none, removing it from the body. Water and solutes will then be added to the blood if necessary by diffusion from the dialysis fluid which runs across the semipermeable tubules containing the circulating blood. The blood is then run through an air bubble trap before it is reintroduced to the vein.

The Effects of Toxicants on Biological Organisms

The problem is to determine whether exposures have occurred before substantial symptoms are present. This is accomplished by a variety of medical tests. The results from these tests must be compared to a medical baseline study, performed before any exposure. Many chemical companies perform baseline studies on new employees prior to employment. Respiratory problems are diagnosed using a spirometer. The patient exhales as hard and as hard as possible into the device. The spirometer measures (1) the total volume exhaled called the forced vital capacity (FVC) with units in litres per second, (2) the forced expiratory flow in the middle range of the vital capacity (FEV 25-75%) in litres per second, and (4) the ratio of the observed (FEV)1 to FVC x 100 (FEV1/FVC%).

Reductions in expiration flow rate are indicative of bronchial disease such as asthma or bronchitis. Reductions in FVC are due to reduction in the lung or chest volume, possibly as a result of fibrosis (an increase in the interstitial fibrous tissue in the lung). The air remaining in the lung after exhalation is called the residual volume (RV). An increase in the RV is indicative of deterioration of the alveoli, possibly due to emphysema. The RV measurement requires a specialised tracer test with helium.

Various Responses to Toxicants

- Effects that are irreversible
- Carcinogen causes cancer
- Mutagen causes chromosome damage
- Reproductive hazard causes damage to reproductive system
- Teratogen causes birth defects
- Effects that may or may not be reversible
- Dermatotoxic affects skin
- Hemotoxic affects blood
- Hepatotoxic affects liver
- Nephrotoxic affects kidneys
- Neurotoxic affects nervous system
- Pulmonotoxic affects lungs

BIOACCUMULATION

Bioaccumulation refers to the accumulation of substances, such as pesticides, or other organic chemicals in an organism. Bioaccumulation occurs when an organism absorbs a toxic substance at a rate greater than that at which the substance is lost. Thus, the longer the biological half life of the substance the greater the risk of chronic poisoning, even if environmental levels of the toxin are not very high. Bioaccumulation, for example in fish, can be predicted by models. Hypotheses for molecular size cutoff criteria for use as bioaccumulation potential indicators are not supported by data. Biotransformation can strongly modify bioaccumulation of chemicals in an organism.

Bioconcentration is a related but more specific term, referring to uptake and accumulation of a substance from water alone. By contrast, bioaccumulation refers to uptake from all sources combined (*e.g.* water, food, air, *etc.*)

Examples

An example of poisoning in the workplace can be seen from the phrase "as mad as a hatter". The process for stiffening the felt used in making hats involved-mercury, which forms organic species such as methylmercury, which is lipid soluble, and tends to accumulate in the brain resulting in mercury poisoning. Other lipid (fat) soluble poisons include tetraethyllead compounds (the lead in leaded petrol), and DDT. These compounds are stored in the body's fat, and when the fatty tissues are used for energy, the compounds are released and cause acute poisoning.

Strontium-90, part of the fallout from atomic bombs, is chemically similar enough to calcium that it is utilized in osteogenesis, where its radiation can cause damage for a long time.

Naturally produced toxins can also bioaccumulate. The marine algal blooms known as "red tides" can result in local filter feeding organisms such as musselsand oysters becoming toxic; coral fish can be responsible for the poisoning known as ciguatera when they accumulate a toxin called ciguatoxin from reef algae.

Some animal species exhibit bioaccumulation as a mode of defense; by consuming toxic plants or animal prey, a species may accumulate the toxin which then presents a deterrent to a potential predator. One example is the tobacco hornworm, which concentrates nicotine to a toxic level in its body as it consumes tobacco plants. Poisoning of small consumers can be passed along the food chain to affect the consumers later on. Other compounds that are not normally considered toxic can be accumulated to toxic levels in organisms. The classic example is of Vitamin A, which becomes concentrated incarnivore livers of *e.g.* polar bears: as a pure carnivore that feeds on other carnivores (seals), they accumulate extremely large amounts of Vitamin A in their livers. It was known by the native peoples of the Arctic that the livers of carnivores should not be eaten, but Arctic explorers have suffered Hypervitaminosis Afrom eating the bear livers (and there has been at least one example of similar poisoning of Antarctic explorers eating husky dog livers). One notable example of this is the expedition of Sir Douglas Mawson, where his exploration companion died from eating the liver of one of their dogs.

Coastal fish (such as the smooth toadfish) and seabirds (such as the Atlantic Puffin) are often monitored for heavy metal bioaccumulation.

In some eutrophic aquatic systems, biodilution can occur. This trend is a decrease in a contaminant with an increase in trophic level and is due to higher concentrations of algae and bacteria to "dilute" the concentration of the pollutant.

How Toxicants are Eliminated from Biological Organisms

Detoxification (**detox** for short) is the physiological or medicinal removal of toxic substances from a living organism, including, but not limited to, the human body and additionally can refer to the period of withdrawal during which an organism returns to homeostasis after long-term use of an addictivesubstance. In medicine, detoxification can be achieved by decontamination of poison ingestion and the use of antidotes as well as techniques such asdialysis and (in a limited number of cases) chelation therapy.

Many alternative medicine practitioners promote various types of detoxification such as detoxification diets. Scientists have described these as a "waste of time and money".[5][6] Sense About Science, a UK-based charitable trust determined that most commercial products' "detox" claims lack any supporting evidence

Types of Detoxification

Alcohol Detoxification

Alcohol detoxification is a process by which a heavy drinker's system is brought back to normal after being used to having alcohol in the body on a continual basis. Serious alcohol addiction results in a decrease in production of GABA, a reuptake inhibitor, because alcohol acts to replace it. Precipitouswithdrawal from long-term alcohol addiction without medical management can cause severe health problems and can be fatal. Alcohol detox is not a treatment for alcoholism. After detoxification, other treatments must be undergone to deal with the underlying addiction that caused the alcohol use.

Drug Detoxification

Drug detoxification is used to reduce or relieve withdrawal symptoms while helping the addicted individual adjust to living without drug use; drug detoxification is not meant to treat addiction but rather an early step in long-term treatment. Detoxification may be achieved drug free or may use medications as an aspect of treatment. Often drug detoxification and treatment will occur in a community program that lasts several months and takes place in a residential rather than medical center.

Drug detoxification varies depending on the location of treatment, but most detox centers provide treatment to avoid the symptoms of physical withdrawal to alcohol & other drugs. Most also incorporate counseling and therapy during detox to help with the consequences of withdrawal.

Metabolic Detoxification

An animal's metabolism can produce harmful substances which it can then make less toxic through reduction, oxidation (collectivelyknown as redoxreactions), conjugation and excretion of molecules from cells or tissues. This is called xenobiotic metabolism. Enzymes that are important in detoxification metabolism include cytochrome P450 oxidases, UDP-glucuronosyltransferases, and glutathione S-transferases. These processes are particularly well-studied as part of drug metabolism, as they influence the pharmacokinetics of a drug in the body.

Alternative Medicine

Certain approaches in alternative medicine claim to remove "toxins" from the body through herbal, electrical or electromagnetic treatments (such as theAqua Detox treatment). These toxins are undefined and have no scientific basis,[6] making the validity of such techniques questionable. There is no evidence for toxic accumulation in these cases,[6] as the liver and kidneys automatically detoxify and excrete many toxic materials including metabolicwastes. Under this theory if toxins are too rapidly released without being safely eliminated (such as metabolizing fat that stores toxins) they can damage the body and cause malaise. Therapies include contrast showers, detoxification foot pads, oil pulling, Gerson therapy, snake-stones, body cleansing,Scientology's Purification Rundown, water fasting, and metabolic therapy.

FACTORS THAT MODIFY TOXICITY

Various characteristics of water and organisms that may change the toxicity of water pollutants.

Abiotic Factors

These are natural characteristics of water that may be influenced by human activity.

i. **Temperature-** temperatures in fresh surface waters of the temperate zone are usually within a range from the freezing point to 30oC. Most aquatic organisms are at the same temperature as the water, and there are upper and lower lethal temperatures beyond which they cannot survive.

Increased water temperature also increases the solubility of many substances, influence the chemical form of some, and governs the amount of oxygen that dissolves in water. Such changes can interact with the direct deleterious effects of elevated temperature. DDT *e.g.* salmonids accumulate more of the pesticide at warm temperature, but following such accumulation there can be near-complete mortality when the fish are subjected to a dropping temperature and their food intake decreases. Presumably DDT is released from storage and becomes actually toxic when body fats are mobilized by the fish at times of low food intake.

There is no single pattern for effects of temperature on toxicity of pollutants to aquatic organisms.

ii. **Dissolved oxygen** - The dissolved oxygen content of fresh water is about 14.6 mg/1 for saturation at 0oC and decreases gradually with temperature to 9.1 mg/1 at 20oC and 7.5 mg/1 at 30oC. Since oxygen is required for respiration, decreased levels can be effective as a limiting factor for aquatic organisms. Environmental or physiological change which affects the rate of respiration flow of a fish will also affect the concentration of poison at the surface of the gill epithelium, and that a known relation exists between these two factors. It also implies that the relation between the increase in toxicity of poisons to fish and reduced dissolved oxygen concentration of the water will be the same for all poisons except those whose toxicities are affected by the PH value of the water.

iii. **Hydrogen ion concentration (pH)-** pH is usually measured as pH, the logarithm of the reciprocal of hydrogen ion activity. Most natural fresh waters are in the range of pH 5-9. Historically, major effects on PH due to human activity have come from acid runoff in coal and metal mining areas. Ionized ammonia (NH4+) has little or no toxicity, but the unionized form (NH3) is quite toxic, the LC50 for salmonids being in the range of 0.2-0.7 mg/l as N. A rise of one pH unit, within the usual middle range of surface waters, increase the proportion of NH3 about sixfold, with a concomitant increase in toxicity.

For some biocides toxicity changes with pH and some it does not. The lethality of a dinitrophenol herbicide was about five times as great to fish at pH 6.9 as at PH 8.0.

iv. **Dissolve salts-** calcium and to a lesser extent, magnesium are the predominant dissolved cations in fresh water and are chiefly responsible for hardness. Atlantic salmon exposed to various pollutants at different salinities. *e.g.* zinc and amonium chloride, tolerance increased from a minimum in freshwater and then dropped in higher salinity.

v. **Binding and sorption-** Suspended and dissolved materials in surface waters may partly detoxify some pollutants. This is true for many metals, particularly copper and some other toxicants. sewage plant effluents and suspended solids have a smaller effect, but colloids, clays, and soil particles can absorb major proportions of copper in water. Between 10 and 80 % of various metals were bound to suspended particles in lake Ontario. Humic and fulvic acids which darken streams in many forested areas, have well-documented effect. Humic acid can cause an order-of-magnitude decrease in copper toxicity in very soft water, but its binding action falls off in harder water.

Biotic Characteristics as Modifying Factors

i. **Test species-** there are real differences among different groups of organisms, and many authors have suggested that pollutants should be tested against an alga and one or more invertebrates, as well as fish to obtain a more complete picture of toxicity. *e.g.* there are some differences even within one group of fish, sunfishes are about 15 times more tolerant of copper than are salmonids and minnows, similar results are also recoded with insecticides and other toxic chemicals.

ii. **Life stage and size-** for aquatic arthropods, the time of molting may be particularly susceptible and could appreciably affect results. For fish, the most sensitive life stage seem to the embryo-larval and early juvenile stages.

Concerning the size, large fish might be expected to be more tolerant of a toxicant, and this is often the case. This is not always true.

iii. **Nutrition, health, and parasitism-** Most descriptions of standard methods for toxicity tests caution that only healthy stocks of fish must be used and that mortality during holding must be low. It would seem likely that any stress of disease would increase or enhance the stress from the toxicant and affect the results.. Adelman and Smith (1976) compared normal goldfish with a stock that had been infested with skin flukes. The latter group had been heavily treated and suffered 50% mortality during holding.

Nutrition of test animals has almost certainly received too little attention as a variable affecting toxicant results. In some cases the effect could be major. Wild populations of the marine copepod showed a sixfold variation in lethal levels of copper, being more resistant when food was more abundant.

Susceptibility to pesticides seems particularly affected by diet. Increased protein in the food was associated with six times greater tolerance of rainbow trout to lethal levels of chlordane.

iv. **Acclimation-** Many hints and incidental observations in the literature suggest that aquatic organisms may acclimate to some toxicants. Much of this information is based on changes in survival times, not threshold lethal concentrations, which would be more meaningful. It might be hypothesized that animals exposed to lethal level of a pollutant could become more tolerant or become weakened, depending on the mode of action of the poison and the type of detoxifying mechanism, if any, available to the animal. A recent series of papers show that both situations occur. Tolerant and susceptible.

Effects of Hazardous Chemicals in Fish (Teratogenesis)

Recently, there has been unprecedented national concern over the problem of hazardous wastes. This concern has been focused upon the adverse environmental and health effects of toxic chemicals from operating hazardous waste treatment, storage and disposal facilities as well as thousands of abandoned waste sites.

Since the late 1950s, it is estimated that more than 750 million tons of toxic chemical wastes have been discarded in 30,000 to 50, 000 hazardous waste sites in the United States. These wastes range from common household trash to complex materials in industrial wastes, sewage sludge, agricultural residue. The disposal of toxic waste creates a major pollution problem and a potential for increased risks to human health.

Optical Malformations

1. Convergence of eyes
2. Synophthalmia
3. Microphthalmia
4. Cyclopia
5. Anophthalmia

The developing optical system is very sensitive, and many investigators have observed optical malformations such as micropthalmia and anopthalmia, as well as cyclopia and intermediate conditions of fusion of the two optical vesicles.

Abnormalities in the development of the optic cups have been observed by many investigators.

Deal (1978) observed disorganized retinas, abnormal pigment distribution, and invase blood sinuses in eyes of medakas treated with methylmercury. Lonning (1977) found reduced pigmentation and protruding lens in oil-treated cod and flatfish embryos. These anomalies were produced when embryos were treated a little as 1-2 hr during cleavage stages.

Microphthalmia and anophthalmia were observed in atlantic silversides and rainbow trout embryos treated with benzo[a]pyrene.

Cause- Reduced mitotic index and increase incidence of pyknotic cells in the optic cups. Tus reduced cell number in the eyes, due to the genotoxic action of the chemical, can account for the abnormally small eyes

Cardiac Malformations

1. Thin atrial and ventricular walls, failure of the heart to bend (tube heart)
2. Tube heart with contractions
3. Tube heart without contractions
4. Pericardial swelling and lack of blood pigment
5. pericardial edema

Cause- The underlying mechanisms in the production of these cardiac abnormalities in fishes have not been ascertained, some work on bird embryos may reveal similar mechanisms. In the literatures mentioned that environmental pollutants (pesticides and insecticides) produce similar effects in bird embryos by lower levels of NAD, therefore by lowering the cells ATP and energy levels. Rogers (1964) demonstrated that the effects of pesticides in chick embryos could be counteracted by adding NAD. It is possible same mechanisms of action are operate in the teleost embryos as well. The looping of the heart has been shown to be caused by shape changes of the hear cells and lower energy levels may not permit this energy requiring process to take place. Many of the toxicants are also inhibitors of cell growth, mitotic inhibitors. This growth inhibition can be responsible for the failure of the heart tube to thicken properly, which, in turn, is responsible for the failure of circulation and subsequent hemostasis. It is likely, that the lack of blood pigment, which is often observed, has a different etiology, involving inhibition of hemoglobin synthesis.

Skeletal Defects

1. Slight bend or kink
2. Major bend or kink
3. Stunted

Cause

Cd -induced vertebral column damage to a decrease in calcium and phosphorus in the bones which weaken them and made them susceptible to curvature by muscle action.

Some pesticides that are neurotoxic may produce skeletal deformities by a physiological rather than developmental mechanism.

Some cases toxicans delay the proliferation and movement of deep cells more strongly than they delay cell differentiation..

Critical Stages V

A number of studies addressed the issue of critical periods in development for the production of anomalies.

1. Early cleavage stage

2. Gastrulation stage

The cleavage and gastrulation stages correspond to the time of induction of the forebrain, defects which are believed to be responsible for convergence of the optic cups. (mercury)

THE DOSE-RESPONSE RELATIONSHIP IS OF PARAMOUNT IMPORTANCE IN TOXICOLOGY

Dose Determines the Biological Response

Dose-Response curve: The relationship between the dose of a chemical (dependent variable) and the response produced (independent variable) follows a predictable pattern. As the dose of a toxicant increases, so does the response, either in terms of the proportion of the population responding or in terms of the severity of the graded responses. For most toxicants, at very low amounts, there will be no detectable effect of the chemical (NOAEL: no observed adverse effect level). In the midrange of doses, the amount of damage will increase as the dose increases (the linear 16-84% of the curve). Larger amounts of chemical will cause increasingly more severe biological responses until a maximum level of damage is reached. Additional toxic effects may also appear along with increased doses, depicting both dose response and dose effect relationships.

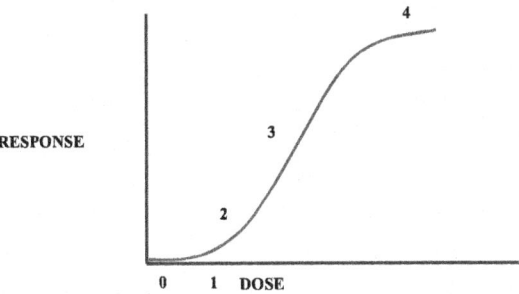

0-1= no adverse effect level

2-3 = linear portion of the curve

4 = maximal response or effect

Quantal responses can be treated as a gradient when data from a population is used. The cumulative proportion of the population responding to a certain dose is plotted for each dose. A similar S-shaped curve is produced since there can be a 10-30 fold variation within a population. If one uses mortality as the response, the dose that is lethal to 50% of the population (LD_{50}) can be calculated from the generated curve. Different toxicants can be compared, and the one with the lowest LD_{50} is the most potent. There are differences in between exposure routes and animals.

Chemical	LD_{50} (mg/kg)	Chemical	LD50 (with route and animal)
Ethyl Alcohol	10,000	Caffeine	620mg/kg – oral mouse 192mg/kg – oral rat 105mg/kg – iv rat 68mg/kg – iv mouse
Sodium Chloride	4,000		
Ferrous Sulfate	1,500	Chlorine (LC 50)	293ppm/1hr – rat 137ppm/1hr – mouse
Morphine Sulfate	900	THC (from marijuana)	175mg/kg – iv mouse 155mg/kg – iv rabbit 100mg/kg – iv dog
Strychnine Sulfate	150		
Nicotine	1	Mercury (I) Chloride	210 mg/kg – oral rat 8 mg/kg – iv mouse
Black Widow	0.55	Mercury (II) Chloride	37 mg/kg – oral rat 10 mg/kg – oral mouse
Curare	0.50	Arsenic acid (V oxidation state)	48 mg/kg – oral rat
Rattle Snake	0.24	Arsenic trioxide (III oxidation state)	20 mg/kg – oral rat
Dioxin (TCDD)	0.001	Dimethylarsenic acid (methylated arsenic form used as a cotton defoliant)	700 mg/kg – oral rat
Botulinum toxin	0.0001		

EXPOSURE

An organism must be exposed to an agent before there is a risk. The physical properties of the chemical and the concentration of the chemical in the environment are important in determining the extent of the exposure.

Toxic effects in a biological system are not produced unless the agent or its metabolic breakdown (biotransformation) products reach appropriate target sites in the body at a concentration and for a length of time sufficient to cause toxicity. We need to define, HOW MUCH, HOW LONG, and HOW OFTEN.

Doses of Chemicals:

Dose = Amount / Animal Mass   *i.e.* mg/kg of animal body weight

A given dose can be compared across animal species

Example: Need to administer 100mg/kg dose of the drug to a mouse, a rat, and a human.

20g mouse would get **2mg** of drug

200g rat would get **20mg** of drug

70kg human would get **7g** of the drug

Duration of Exposure:

Acute:	< 24hr	usually single exposure
Subacute	1 month	repeated doses
Subchronic	1-3 months	repeated doses
Chronic	>3months	repeated doses

Toxic Effects: Acute *vs* Chronic

Overtime, the amount of chemical in the body can build up, it can redistribute, or it can overwhelm repair mechanisms.

Toxicologists are most interested in what is the most common scenario, chronic exposure to low doses.

	Single Dose	Repeated Dose
Benzene	CNS Depression	Leukemia

Frequency of Exposure:

The frequency of the exposure affects the concentration at the target site — can build up to a steady level--why some medications are taken three times a day *vs.* once a day to give the wanted effect.

Route and Site of Exposure:

Ingestion (gastrointestinal tract)

Inhalation (lungs)

Dermal / topical (skin)

Parenteral (intravenous--iv, intramuscular — im, intraperitoneal — ip)

Typical Effectiveness of Route of Exposure

iv > inhalation > ip > im > ingestion > topical

ABSORPTION, DISTRIBUTION, METABOLISM, AND EXCRETION OF TOXICANTS (ADME)

The toxicant may have to pass many barriers to get to its site of action

Absorption:

Absorption--the ability of a chemical agent to enter the blood.

Similar blood levels are more likely to give similar effects than similar administered doses. Blood is in equilibrium with the other tissues and target sites.

Intravenous	No limiting factors in absorption (100% bioavailable)
Inhalation	Must penetrate alveolar sacs of lungs but then into capillary bed
Ingestion	Requires absorption through GI tract and is subject to 1st pass effect
Intraperitoneal	Like ingestion (still 1st pass effect) but does not require absorption through the GI tract
Dermal/Topical	Requires absorption through the skin

Distribution:

Distribution — the process in which a chemical agent translocates throughout the body. The blood carries the agent to and from its site of action, storage depots, organs of biotransformation, and organs of elimination. The rate of distribution is usually rapid, and is determined primarily by blood flow and the chemical characteristics of the toxicant (its affinity for the tissue and the partition coefficient). The distribution of a chemical may change over time.

Storage — DDT in Fatty tissues

Lead and Fluoride in Bone

Metabolism:

Metabolism (biotransformation) — the process by which administered chemicals (parent compounds) are modified by the organism, usually *via* enzymes. The primary objective of metabolism is to make chemical agents more water soluble and easier to excrete by

Decreasing lipid solubility → Decreased amount that reaches target

Increasing ionization → Increased rate of excretion → Decrease toxicity

In some situations, biotransformation results in the formation of reactive metabolites — Bioactivation.

Whether it is the parent compound or the metabolite, it is the active compound that does the damage.

Excretion:

Toxicants are eliminated from the body by several routes.

Urinary excretion

Water soluble products are filtered out of the blood and excreted into the urine.

Exhalation

Volatile compounds are exhaled through breathing

Biliary Excretion *via* Fecal Excretion

Compounds can be extracted by the liver, biotransformed, and excreted into the bile. The bile drains into the small intestine where the eliminated compound can be excreted into the feces.

Fecal excretion also rids the body of non-absorbed compounds which pass through the GI tract.

MORE ON METABOLISM

Biotransformation can occur at any point during the compound's trek from absorption to excretion.

Biotransformation can drastically effect the rate of clearance of compounds

Without Biotransformation		With Biotransformation
Ethanol	4 weeks	10mL/hr
Phenobarbital	5 months	8hr
DDT	infinity	days to weeks

Key organs of Biotransformation:

LIVER (High)

Lung, Kidney, Intestine (Medium)

Others (Low)

Biotransformation Pathways

Phase I enzymes: Makes the toxicant more soluble

Phase II Enzymes: Links with a soluble agent (conjugation)

Individual Susceptibility:

Individual variation of the organism will affect the absorption, distribution, metabolism, and excretion of the toxicant, and there fore the effect of the toxicant.

There can be a 10-30 fold difference in response to a toxicant in a population due to:

Genetics — species, strain variations, inter-individual variations

Gender

Age (young and old)

Nutritional status

Health conditions

Previous or concurrent exposure to other substances

Relativity

Toxicants are compared for relative toxicity based on the LD, ED, or TD curves. If the response – dose curve for chemical A is to the right of the response – dose curve for chemical B, then chemical A is more toxic. Care must be taken when comparing two responses – dose curves when partial data are available. If the slopes of the curves differ substantially, the situation shown in Figure 5.1 might occur. If only a single data point is available in the upper part of the curves, it might appear that chemical A is always more toxic than chemical B. The complete data shows that chemical B is more toxic at lower doses.

THRESHOLD LIMIT VALUES

The lowest value on the response versus dose curve is called the threshold dose. Below this dose the body is able to detoxify and eliminate the agent without any detectable effects. In reality the response is only identically zero when the dose is zero but for small doses the response is not detectable.

The American Conference of Governmental Industrial Hygienists (ACGIH) has established threshold doses called threshold limit values (TLVs) for a large number of chemical agents. The TLV refers to an airborne concentration that corresponds to conditions where no adverse effects are normally expected during a worker's lifetime. The exposure occurs only during normal working hours, eight hours per day and five days per week. The TLV was formerly called the maximum allowable concentration (MAC). There are three different types of TLVs (TLV-TWA, TLV-STEL and TLV-C) with precise definitions provided in Table. More TLV-TWA data is available than TWA-STEL or TLV-C data.

The United States Occupational Safety and Health Administration (OSHA) has defined their own threshold dose called a permissible exposure level (PEL). The PEL values follow the TLV-TWA of the ACGIH very closely. However, the PEL values are not as numerous and are not updated as frequently. The TLV values are often somewhat more conservative.

Dose *versus* Response

The science of toxicology is based on the principle that there is a relationship between a toxic reaction (the response) and the amount of poison received (the dose). An important assumption in this relationship is that there is almost always a dose below which no response occurs or can be measured. A second assumption is that once a maximum response is reached any further increases in the dose will not result in any increased effect.

One particular instance in which this dose-response relationship does not hold true, is in regard to true allergic reactions. Allergic reactions are special kinds of changes in the immune system; they are not really toxic responses. The difference between allergies and toxic reactions is that a toxic effect is directly the result of the toxic chemical acting on cells. Allergic responses are the result of a chemical stimulating the body to release natural chemicals which are in turn directly responsible for the effects seen. Thus, in an allergic reaction, the chemical acts merely as a trigger, not as the bullet.

For all other types of toxicity, knowing the dose-response relationship is a necessary part of understanding the cause and effect relationship between chemical exposure and illness. As Paracelsus once wrote, "The right dose differentiates a poison from a remedy." Keep in mind that the toxicity of a chemical is an inherent quality of the chemical and cannot be changed without changing the chemical to another form. The toxic effects on an organism are related to the amount of exposure.

Measures of Exposure

Exposure to poisons can be intentional or unintentional. The effects of exposure to poisons vary with the amount of exposure, which is another way of saying "the dose." Usually when we think of dose, we think in terms of taking one vitamin capsule a day or two aspirin every four hours, or something like that. Contamination of food or water with chemicals can also provide doses of chemicals each time we eat or drink. Some commonly used measures for expressing levels of contaminants are listed in table. These measures tell us how much of the chemical is in food, water or air. The amount we eat, drink, or breathe determines the actual dose we receive.

Concentrations of chemicals in the environment are most commonly expressed as ppm and ppb. Government tolerance limits for various poisons usually use these abbreviations. Remember that these are extremely small quantities. For example, if you put one teaspoon of salt in two gallons of water the resulting salt concentration would be approximately 1,000 ppm and it would not even taste salty!

Table. Measurements for Expressing Levels of Contaminants in Food and Water.

Dose	Abbrev.	Metric equivalent	Abbrev.	Approx. amt. in water
parts per million	ppm	milligrams per kilogram	mg/kg	1 teaspoon per 1,000 gallons
parts per billion	ppb	micrograms per kilogram	ug/kg	1 teaspoon per 1,000,000 gallons

Dose-Effect Relationships

The dose of a poison is going to determine the degree of effect it produces. The following example illustrates this principle. Suppose ten goldfish are in a ten-gallon tank and we add one ounce of 100-proof whiskey to the water every five minutes until all the fish get drunk and swim upside down. Probably none would swim upside down after the first two or three shots. After four or five, a very sensitive fish might. After six or eight shots another one or two might. With a dose of ten shots, five of the ten fish might be swimming upside down. After fifteen shots, there might be only one fish swimming properly and it too would turn over after seventeen or eighteen shots.

The effect measured in this example is swimming upside down. Individual sensitivity to alcohol varies, as does individual sensitivity to other poisons. There is a dose level at which none of the fish swim upside down (no observed effect). There is also a dose level at which all of the fish swim upside down. The dose level at which 50 percent of the fish have turned over is known as the ED50, which means effective dose for 50 percent of the fish tested. The ED50 of any poison varies depending on the effect measured. In general, the less severe the effect measured, the lower the ED50 for that particular effect. Obviously poisons are not tested in humans in such a fashion. Instead, animals are used to predict the toxicity that may occur in humans.

One of the more commonly used measures of toxicity is the LD50. The LD50 (the lethal dose for 50 percent of the animals tested) of a poison is usually expressed in milligrams of chemical per kilogram of body weight (mg/kg). A chemical with a small LD50 (like 5 mg/kg) is very highly toxic. A chemical with a large LD50 (1,000 to 5,000 mg/kg) is practically non-toxic. The LD50 says nothing about non-lethal toxic effects though. A chemical may have a large LD50, but may produce illness at very small exposure levels. It is incorrect to say that chemicals with small LD50s are more dangerous than chemicals with large LD50s, they are simply more toxic. The danger, or risk of adverse effect of chemicals, is mostly determined by how they are used, not by the inherent toxicity of the chemical itself.

The LD50s of different poisons may be easily compared; however, it is always necessary to know which species was used for the tests and how the poison was administered (the route of exposure), since the LD50 of a poison may vary considerably based on the species of animal and the way exposure occurs. Some poisons may be extremely toxic if swallowed (oral exposure) and not very toxic at all if splashed on the skin (dermal exposure). If the oral LD50 of a poison were 10 mg/kg, 50 percent of the animals who swallowed 10 mg/kg would be expected to die and 50 percent to live. The LD50 is determined mathematically, and in actual tests using the LD50, it would be unusual to get an exact 50% response. One test might produce 30% mortality and another might produce 70% mortality. Averaged out over many tests, the numbers would approach 50%, if the original LD50 determination was valid.

The potency of a poison is a measure of its strength compared to other poisons. The more potent the poison, the less it takes to kill; the less potent the poison, the more it takes to kill. The potencies of poisons are often compared using signal words.

The designation toxic dose (TD) is used to indicate the dose (exposure) that will produce signs of toxicity in a certain percentage of animals. The TD50 is the toxic dose for 50 percent of the animals tested. The larger the TD the more poison it takes to produce signs of toxicity. The toxic dose does not give any information about the lethal dose because toxic effects (for example, nausea and vomiting) may not be directly related to the way that the chemical causes death. The toxicity of a chemical is an inherent property of the chemical itself. It is also true that chemicals can cause different types of toxic effects, at different dose levels, depending on the animal species tested. For this reason, when using the toxic dose designation it is useful to precisely define the type of toxicity measured, the animal species tested, and the dose and route of administration.

Table. Toxicity Rating Scale and Labeling Requirements for Pesticides.

Category	Signal word required on label	LD50 oral mg/kg(ppm)	LD50 dermal mg/kg(ppm)	Probable oral lethal dose
I highly toxic	DANGER-POISON (skull and cross-bones)	less than 50	less than 200	a few drops to a teaspoon
II moderately toxic	WARNING	51 to 500	200 to 2,000	over 1 teaspoon to 1 ounce
III slightly toxic	CAUTION	over 500	over 2,000	over 1 ounce
IV practically non-toxic	none required			

Toxicity assessment is quite complex, many factors can affect the results of toxicity tests. Some of these factors include variables like temperature, food, light, and stressful environmental conditions. Other factors related to the animal itself include age, sex, health, and hormonal status.

The NOEL (no observable effect level) is the highest dose or exposure level of a poison that produces no noticeable toxic effect on animals. From our previous fish example, we know that there is a dose below which no effect is seen. In toxicology, residue tolerance levels of poisons that are permitted in food or in drinking water, for instance, are usually set from 100 to 1,000 times less than the NOEL to provide a wide margin of safety for humans.

The TLV (threshold limit value) for a chemical is the airborne concentration of the chemical (expressed in ppm) that produces no adverse effects in workers exposed for eight hours per day five days per week. The TLV is usually set to prevent minor toxic effects like skin or eye irritation.

Very often people compare poisons based on their LD50's and base decisions about the safety of a chemical based on this number. This is an over-simplified approach to comparing chemicals because the LD50 is simply one point on the dose-response curve that reflects the potential of the compound to cause death. What is more important in assessing chemical safety is the threshold dose, and the slope of the dose-response curve, which shows how fast the response increases as the dose increases. Which of these chemicals is more toxic? Answer this question for doses below the LD50 and it is chemical A which is more toxic, at the LD50 they are the same, and above the LD50, chemical B is more toxic. While the LD50 can provide some useful information, it is of limited value in risk assessment because the LD50 only reflects information about the lethal effects of the chemical. It is quite possible that a chemical will produce a very undesirable toxic effect (such as reproductive toxicity or birth defects) at doses which cause no deaths at all.

A true assessment of chemical toxicity involves comparisons of numerous dose-response curves covering many different types of toxic effects. The determination of which pesticides will be Restricted Use Pesticides involves this approach.

Some Restricted Use Pesticides have very large LD50s (low acute oral toxicity), however, they may be very strong skin or eye irritants and thus require special handling.

The knowledge gained from dose-response studies in animals is used to set standards for human exposure and the amount of chemical residue that is allowed in the environment. As mentioned previously, numerous dose-response relationships must be determined, in many different species. Without this information, it is impossible to accurately predict the health risks associated with chemical exposure. With adequate information, we can make informed decisions about chemical exposure and work to minimize the risk to human health and the environment.

Dose–response relationship

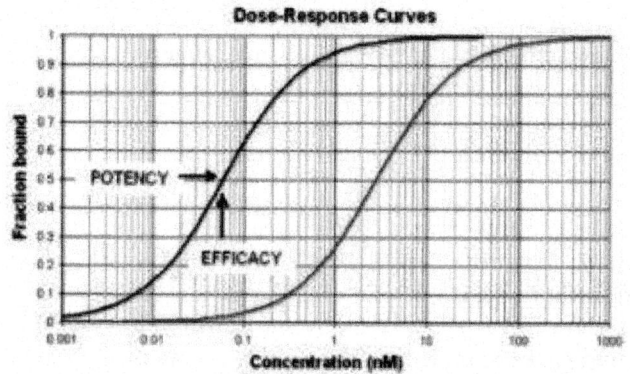

Semi-log plots of two agonists with different K_d; the blue curve indicates a more potent agonist than the green curve (higher response for the same dose).

The **dose–response relationship**, or **exposure–response relationship**, describes the change in effect on an organism caused by differing levels of exposure (or doses) to a stressor (usually a chemical) after a certain exposure time. This may apply to individuals (*e.g.*: a small amount has no significant effect, a large amount is fatal), or to populations (*e.g.*: how many people or organisms are affected at different levels of exposure).

Studying dose response, and developing dose–response models, is central to determining "safe" and "hazardous" levels and dosages for drugs, potential pollutants, and other substances to which humans or other organisms are exposed. These conclusions are often the basis for public policy. The U.S. Environmental Protection Agency has developed extensive guidance and reports on dose-response modeling and assessment, as well as software.

Dose-response relationships generally depend on the exposure time and exposure route (*e.g.*, inhalation, dietary intake); quantifying the response after a different exposure time or for a different route leads to a different relationship and possibly different conclusions on the effects of the stressor under consideration. This limitation is caused by the complexity of biological systems and the often unknown biological processes operating between the external exposure and the adverse cellular or tissue response.

Dose–response Curve

A **dose–response curve** is a simple X–Y graph relating the magnitude of a stressor (*e.g.* concentration of a pollutant, amount of a drug, temperature, intensity of radiation) to the response of the receptor (*e.g.* organism under study). The response may be a physiological or biochemical response, or even death (mortality), and thus can be counts (or proportion, *e.g.*, mortality rate), ordered descriptive categories (*e.g.*, severity of a lesion), or continuous measurements (*e.g.*, blood pressure). A number of effects (or endpoints) can be studied, often at different organizational levels (*e.g.*, population, whole animal, tissue, cell).

The measured dose (usually in milligrams, micrograms, or grams per kilogram of body-weight for oral exposures or milligrams per cubic meter of ambient air for inhalation exposures) is generally plotted on the X axis and the response is plotted on the Y axis. Other dose units include moles per body-weight, moles per animal, and for dermal exposure, moles per square centimeter. In some cases, it is the logarithm of the dose that is plotted on the X axis, and in such cases the curve is typically sigmoidal, with the steepest portion in the middle. Biologically based models using dose are preferred over the use of log(dose) because the latter can visually imply a threshold dose when in fact there is none.

The first point along the graph where a response above zero (or above the control response) is reached is usually referred to as a threshold-dose. For most beneficial or recreational drugs, the desired effects are found at doses slightly greater than the threshold dose. At higher doses, undesired side effects appear and grow stronger as the dose increases. The more potent a particular substance is, the steeper this curve will be. In quantitative situations, the Y-axis often is designated by percentages, which refer to the percentage of exposed individuals registering a standard response (which may be death, as in LD_{50}). Such a curve is referred to as a quantal dose-response curve, distinguishing it from a graded dose-response curve, where response is continuous (either measured, or by judgment).

A commonly used dose-response curve is the EC_{50} curve, the **half maximal effective concentration**, where the EC_{50} point is defined as the inflection point of the curve.

Statistical analysis of dose-response curves may be performed by regression methods such as the probit model or logit model, or other methods such as the Spearman-Karber method. Empirical models based on nonlinear regression are usually preferred over the use of some transformation of the data that linearizes the dose-response relationship.

Dose–response curves can be fit to the Hill equation (biochemistry) to determine cooperativity.

Problems with Linear Model

The concept of linear dose-response relationship, thresholds, and all-or-nothing responses may not apply to non-linear situations. A threshold model or linear no-threshold model may be more appropriate, depending on the circumstances. A recent critique[5] of these models as they apply to endocrine disruptors argues for a substantial revision of testing and toxicological models at low doses.

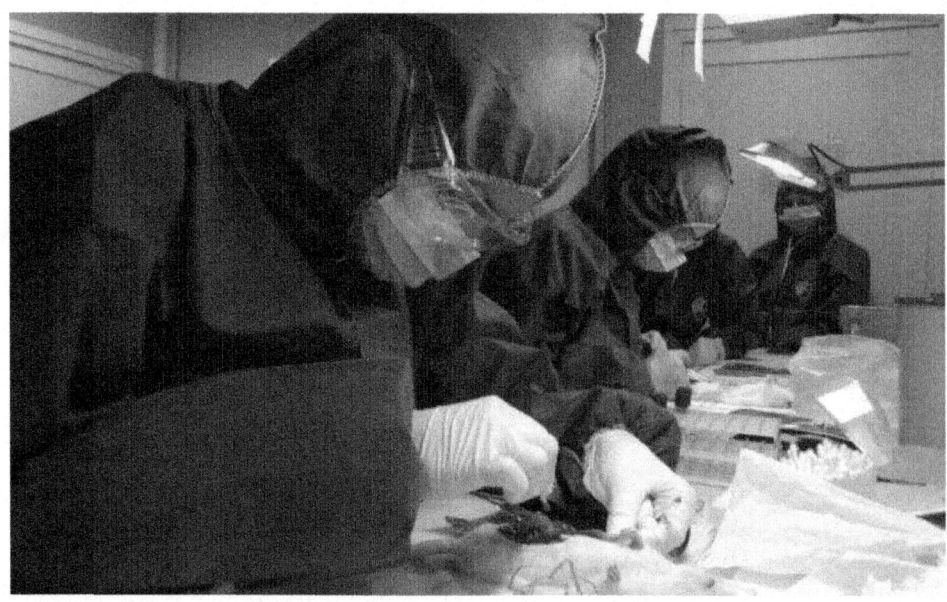

Figure : Chemical Toxicology

Chapter 3

INDUSTRIAL HYGIENE IN CHEMICAL PROCESS

Industrial Hygiene (also referred to as occupational hygiene) has been defined as the science of protecting and enhancing the health and safety of people at work and in their communities (ABIH 2010). Best practice can be defined as the best means to achieve a desired goal (*i.e.*, health and safety). There are only a few technical references that identify approaches, tools, or methods as being a best practice. This chapter summarizes these best practices within the context of five recognized aspects of IH (see Figure 1):

- anticipation
- recognition
- evaluation
- prevention
- control

One of the major responsibilities of the industrial hygienist is to anticipate, identify, and solve potential health problems within plants. Chemical process technology, however, is so complex that this task requires the concerted efforts of industrial hygienists, process designers, opera- tors, laboratory personnel, and management. The industrial hygienist helps the effectiveness of the overall program by working with these plant personnel. For these reasons industrial hygiene (particularly identification) must be a part of the education process of chemists, engineers, and managers. Many hazardous chemicals are handled safely on a daily basis within chemical plants. To achieve this operating success, all potential hazards must be identified and controlled. When toxic and/or flammable chemicals are handled, the potentially hazardous conditions may be numerous — in large plants there may be thousands. To be safe under these conditions requires discipline, skill, concern, and attention to detail.

Occupational (or "industrial" in the U.S.) hygiene (IH) is the control and prevention of hazards from work that may result in injury, illness, or affect the well being of workers. These hazards or stressors are typically divided into the categories biological, chemical physical, ergonomic and psycosocial The risk of a health effect from a given stressor is a function of the hazard multiplied by the exposure to the individual or group for chemicals, the hazard can be understood by the dose response profile most often based on toxicological studies or models. Occupational hygienists work closely with toxicologists (see Toxicology) for understanding chemical hazards, physicists (see Physics) for physical hazards, and physicians and microbiologists for biological hazards (see Microbiological tropical medicine Infection)

The British Occupational Hygiene Society (BOHS) defines that "occupational hygiene is about the prevention of ill-health from work, through recognizing, evaluating and controlling the risks". The International Occupational Hygiene Association (IOHA) refers to occupational hygiene *as the discipline of anticipating, recognizing, evaluating and controlling health hazards in the working environment with the objective of protecting worker health and well-being and safeguarding the community at large.* The term "occupational hygiene" (used in the UK and Commonwealth countries as well as much of Europe) is synonymous with *industrial hygiene* (used in the US, Latin America, and other countries that received initial technical support or training from US sources). The term "industrial hygiene" traditionally stems from industries with construction, mining or manufacturing and "occupational hygiene" refers to all types of industry such as those listed for "industrial hygiene" as well as financial and support services industries and refers to "work", "workplace" and "place of work" in general. *Environmental hygiene* addresses similar issues to *occupational hygiene*, but is likely to be about broad industry or broad issues affecting the local community, broader society, region or country.

The profession of **occupational hygiene** uses strict and rigorous scientific methodology and often requires professional experience in determining the potential for hazard and exposure risks in workplace and environmental studies. This aspect of occupational hygiene is often referred to as the "art" of occupational hygiene and is used in a similar sense to the "art" of medicine. In fact "occupational hygiene" is both an aspect of preventive medicine and in particular occupational medicine, in that its goal is to prevent industrial disease, using the science of risk management, exposure assesment and industrial safety. Ultimately professionals seek to implement "safe" systems, procedures or methods to be applied in the workplace or to the environment.

THE SOCIAL ROLE OF OCCUPATIONAL HYGIENE

Occupational hygienists have been involved historically with changing the perception of society about the nature and extent of hazards and preventing exposures in the workplace and communities. Many occupational hygienists work day-to-day with industrial situations that require control or improvement to the workplace situation however larger social issues affecting whole industries have occurred in the past *e.g.* since 1900, asbestos exposures that have affected the lives of tens of thousands of people. Occupational hygienists have become more engaged in understanding and managing exposure risks to consumers from products with new regulations such as REACh (Registration, Evaluation, Authorization and Restriction of Chemicals).

More recent issues affecting broader society are, for example in 1976, legionnaires' disease or legionellesis. More recently again in the 1990s radon and in the 2000s the effects of mold from indoor air quality situations in the home and at work. In the later part of the 2000s concern has been raised about the health effects of nanoparticles.

Many of these issues have required the coordination over a number of years of a number of medical and para professionals in detecting and then characterizing the nature of the issue, both in terms of the hazard and in terms of the risk to the workplace and ultimately to society. This has involved occupational hygienists in research, collection of data and to develop suitable and satisfactory control methodologies.

WORKPLACE ASSESSMENT METHODS

Although there are many aspects to occupational hygiene work the most known and sought after is in determining or estimating potential or actual to-expesoures to hazards. For many chemicals and physical hazards, occupational exposure limits have been derived using toxicological, epidemiological and medical data allowing hygienists to reduce the risks of health effects by implementing the "Hierarchy of Hazard Controls".

Several methods can be applied in assessing the workplace or environment for exposure to a known or suspected hazard. Occupational hygienists do not rely on the accuracy of the equipment or method used but in knowing with certainty and precision the limits of the equipment or method being used and the error or variance given by using that particular equipment or method. Well known methods for performing occupational exposure assessments can be found in "A Strategy for Assessing and Managing Occupational Exposures, Third Edition Edited by Joselito S. Ignacio and William H. Bullock".

The main steps outlined for assessing and managing occupational exposures:
- Basic Characterization (identify agents, hazards, people potentially exposed and existing exposure controls)
- Exposure Assessment (select occupational exposure limits, hazard bands, relevant toxicological data to determine if exposures are "acceptable", "unacceptable" or "uncertain")
- Exposure Controls (for "unacceptable" or "uncertain" exposures)
- Further Information Gathering (for "uncertain" exposures)
- Hazard Communication (for all exposures)
- Reassessment (as needed) / Management of Change

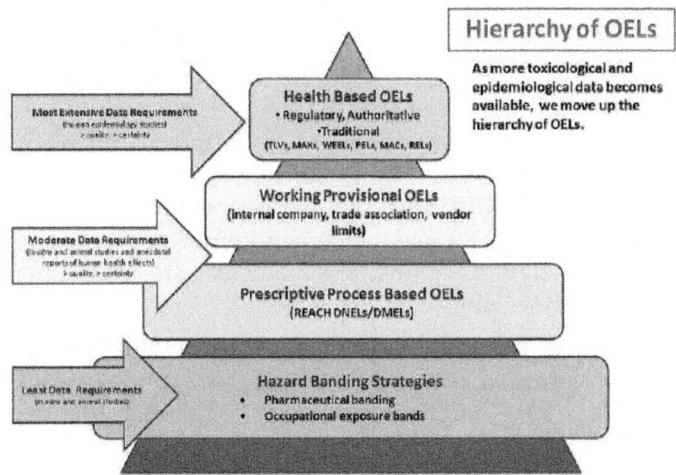

Figure : Hierarchy of occupational exposure limits (OELs)

Basic Characterization & Walk Through Surveys

The first step in understanding health risks related to exposures requires the collection of "basic characterization" information from available sources. A traditional method applied by occupational hygienists to initially survey a workplace or environment is used to determine both the types and possible exposures from hazards (*e.g.* noise, chemicals, radiation). The **walk-through survey** can be targeted or limited to particular hazards such as silica dust, or noise, to focus attention on control of all hazards to workers. A full walk-through survey is frequently used to provide information on establishing a framework for future investigations, prioritizing hazards, determining the requirements for measurement and establishing some immediate control of potential exposures. The Health Hazard Evaluation Program from the National Institute for Occupational Safety and Health is an example of an industrial hygiene walk-through survey. Other sources of basic characterization information include worker interviews, observing exposure tasks, material safety data sheets, workforce scheduling, production data, equipment and maintenance schedules to identify potential exposure agents and people possibly exposed.

Sampling Survey Equipment

An occupational hygienist may use one or a number of commercially available electronic measuring devices to measure noise, vibration, ionizing and non-ionizing radiation, dust, solvents, gases, etcetera. Each device is often specifically designed to measure a specific or particular type of contaminant. Such devices are often subject to multiple interferences. Electronic devices need to be calibrated before and after use to ensure the accuracy of the measurements taken and often require a system of certifying the precision of the instrument.

Dust Sampling

Nuisance dust is considered to be the total dust in air including inhalable and respirable fractions.

Various dust sampling methods exist that are internationally recognised. **Inhalable** dust is determined using the modern equivalent of the Institute of Occupational Medicine (IOM) MRE 113A monitor (see section on workplace exposure, measurement & modelling). Inhalable dust is considered to be dust of less than 100 micrometers aerodynamic equivalent diameter (AED) that enters through the nose and or mouth. See Lungs

Respirable dust is sampled using a cyclone dust sampler design to sample for a specific fraction of dust AED at a set flow rate. The respirable dust fraction is dust that enters the 'deep lung' and is considered to be less than 10 micrometers AED.

Nuisance, inhalable and respirable dust fractions are all sampled using a constant volumetric pump for a specific sampling period. By knowing the mass of the sample collected and the volume of air sampled a concentration for the fraction sampled can be given in milligrams (mg) per metre cubed (m3). From such samples the amount of inhalable or respirable dust can be determined and compared to the relevant occupational exposure limits.

By use of inhalable, respirable or other suitable sampler (7 hole, 5 hole, et cetera) these dust sampling methods can also used to determine metal exposure in the air. This requires collection of the sample on a methyl-cellulose ester (MCE) filter and acid digestion of the collection media in the laboratory followed by measuring metal concentration though an atomic absorption (or emission) spectrophotometery. Both the UK Health and Safety Laboratory and NIOSH Manual of Analytical Methods have specific methodologies for a broad range of metals in air found in industrial processing (smelting, foundries, etcetera).

A further method exists for the determination of asbestos, fibreglass, synthetic mineral fibre and ceramic mineral fibre dust in air. This is the membrane filter method (MFM) and requires the collection of the dust on a grided filter for estimation of exposure by the counting of 'conforming' fibres in 100 fields through a microscope. Results are quantified on the basis of number of fibres per millilitre of air (f/ml). Many countries strictly regulate the methodology applied to the MFM.

Chemical Sampling

Two types of chemically absorbent tubes are used to sample for a wide range of chemical substances. Traditionally a chemical absorbent 'tube' (a glass or stainless steel tube of between 2 and 10 mm internal diameter) filled with very fine absorbent silica (hydrophilic) or carbon, such as coconut charcoal (lypophylic), is used in a sampling line where air is drawn through the absorbent material for between four hours (minimum workplace sample) to 24 hours (environmental sample) period. The hydrophilic material readily absorbs water-soluble chemical and the lypophylic material absorbs non water-soluble materials. The absorbent material is then chemically or physically extracted and measurements performed using various gas chromatograph or mass spectrometry methods. These absorbent tube methods have the advantage of being usable for a wide range of potential contaminates. However, they are relatively expensive methods, are time consuming and require significant expertise in sampling and chemical analysis. A frequent complaint of workers is in having to wear the sampling pump (up to 1 kg) for several days of work to provide adequate data for the required statistical certainty determination of the exposure.

In the last few decades, advances have been made in 'passive' badge technology. These samplers can now be purchased to measure one chemical (*e.g.* formaldehyde) or a chemical type (*e.g.* ketones) or a broad spectrum of chemicals (*e.g.* solvents). They are relatively easy to set up and use. However, considerable cost can still be incurred in analysis of the 'badge'. They weigh 20 to 30 grams and workers do not complain about their presence. Unfortunately 'badges' may not exist for all types of workplace sampling that may be required and the charcoal or silica method may sometimes have to be applied.

From the sampling method, results are expressed in milligrams per cubic meter (mg/m3) or parts per million (PPM) and compared to the relevantoccupational exposure limits.

It is a critical part of the exposure determination that the method of sampling for the specific contaminate exposure is directly linked to the exposure standard used. Many countries regulate both the exposure standard, the method used to determine the exposure and the methods to be used for chemical or other analysis of the samples collected.

**Hierarchies for Effective and Efficient
Protection of Workers & Communities**

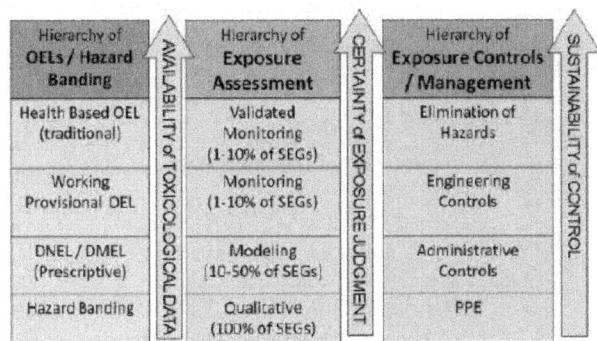

Figure : Simple representation of exposure risk assessment and management
hierarchy based on available information

Exposure Management and Controls

The hierarchy of control defines the approach used to reduce exposure risks protecting workers and communities. These methods include elimination, substitution, engineering controls (isolation or ventilation), administrative controls and personal protective equipment. Occupational hygienists, engineers, maintenance, management and employees should all be consulted for selecting and designing the most effective and efficient controls based on the hierarchy of control.

General Activities

The occupational hygienist may be involved with the assessment and control of physical, chemical, biologicalor environmental hazards in the workplace or community that could cause injury or disease. Physical hazards may include noise, temperature extremes, illumination extremes, ionizing or non-ionizing radiation, and ergonomics. Chemical hazards related to dangerous goods or hazardous substances are frequently investigated by occupational hygienists. Other related areas including indoor air quality (IAQ) and safety may also receive the attention of the occupational hygienist. Biological hazards may stem from the potential for legionella exposure at work or the investigation of biological injury or effects at work, such as dermatitis may be investigated.

As part of the investigation process, the occupational hygienist may be called upon to communicate effectively regarding the nature of the hazard, the potential for risk, and the appropriate methods of control. Appropriate controls are selected from the hierarchy of control: by elimination, substitution, engineering, administration and personal protective equipment (PPE) to control the hazard or eliminate the risk. Such controls may involve recommendations as simple as appropriate PPE such as a 'basic' particulate dust mask to occasionally designing dust extraction ventilation systems, work places or management systems to manage people and programs for the preservation of health and well-being of those who enter a workplace.

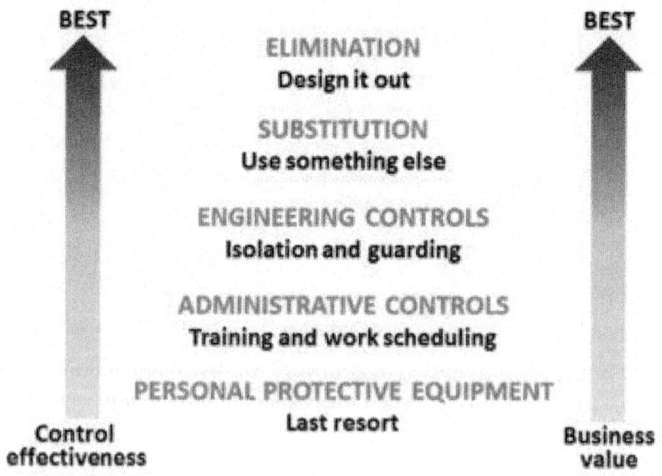

Figure : The hierarchy of controls is an important tool to determine how to control hazards most efficiently and effectively in a workplace.

Education

The basis of the technical knowledge of occupational hygiene is from competent training in the following areas of science and management.

- Basic Sciences (Biology, Chemistry, Mathematics (Statistics), Physics);
- Occupational Diseases (Illness, injury and health surveillance (biostatistics, epidemiology, toxicology));
- Health Hazards (Biological, Chemical and Physical hazards, Ergonomics and Human Factors);
- Working Environments (Mining, Industrial, Manufacturing, transport and storage, service industries and offices);
- Programme Management Principles (professional and business ethics, work site and incident investigation methods, exposure guidelines, Occupational exposure limits, jurisdictional based regulations, hazard identification, risk assessment and risk communication, data management, fire evacuation and other emergency responses);
- Sampling, measurement and evaluation practices (instrumentation, sampling protocols, methods or techniques, analytical chemistry);
- Hazard Controls (elimination, substitution, engineering, administrative, PPE and Air Conditioning and Extraction Ventilation);
- Environment (air pollution, hazardous waste).

However, it is not rote knowledge that identifies a competent occupational hygienist. There is an "art" to applying the technical principles in a manner that provides a reasonable solution for workplace and environmental issues. In effect an experienced "mentor", who has experience in occupational hygiene is required to show a new occupational hygienist how to apply the learned scientific and management knowledge in the workplace and to the environment issue to satisfactorily resolve the problem.

To be a professional occupational hygienist, experience in as wide a practice as possible is required to demonstrate knowledge in areas of occupational hygiene. This is difficult for "specialists" or those who practice in narrow subject areas. Limiting experience to individual subject like asbestos remediation, confined spaces, indoor air quality, or lead abatement, or learning only through a textbook or "review course" can be a disadvantage when required to demonstrate competence in other areas of occupational hygiene.

Information presented in Wikipedia can be considered to be only an outline of the requirements for professional occupational hygiene training. This is because the actual requirements in any country, state or region may vary due to educational resources available, industry demand or regulatory mandated requirements.

During 2010, the Occupational Hygiene Training Association (OHTA) through sponsorship provided by the IOHA initiated a training scheme for those with an interest in or those requiring training in occupational hygiene. These training modules can be downloaded and used freely. The available subject modules (Basic Principles in Occupational Hygiene, Health Effects of Hazardous Substances, Measurement of Hazardous Substances, Thermal Environment, Noise, Asbestos, Control, Ergonomics) are aimed at the 'foundation' and 'intermediate' levels in Occupational Hygiene. Although the modules can be used freely without supervision attendance at an accredited training course is encouraged. These training modules are available from OH Learning.com

Academic programs offering industrial hygiene Bachelors or Masters degrees in United States may apply to the Accreditation Board for Engineering and Technology (ABET) to have their program accredited. As of October 1, 2006, 27 institutions have accredited their industrial hygiene programs. Accreditation is not available for Doctoral programs.

In the U.S. the training of IH professionals is supported by [National Institute for Occupational Safety and Health]] through their NIOSH Education and Research Centers.

Professional Societies

The development of industrial hygiene societies originated in the United States, beginning with the first convening of members for the American Conference of Governmental Industrial Hygienists in 1938, and the formation of the American Industrial Hygiene Association in 1939. In the United Kingdom, the British Occupational Hygiene Society started in 1953. Through the years, professional occupational societies have formed in many different countries, leading to the formation of the International Occupational Hygiene Association in 1987, in order to promote and develop occupational hygiene worldwide through the member organizations.[8] The IOHA has grown to 29 member organizations, representing over 20,000 occupational hygienists worldwide, with representation from countries present in every continent.[8][9]

Occupational Hygiene Peer-reviewed Literature

There are several academic journals specifically focused on publishing studies and research in the occupational health field. The *Journal of Occupational and Environmental Hygiene* (JOEH) has been published jointly since 2004 by the American Industrial Hygiene Association and the American Conference of Governmental Industrial Hygienists, replacing the former *American Industrial Hygiene Association Journal* and *Applied Occupational & Environmental Hygiene* journals. Another seminal occupational hygiene journal would be *The Annals of Occuapational Hygiene*, published by the British Occupational Hygiene Society since 1958.[11] Further, The National Institute for Occupational Safety and Health maintains a searchable bibliographic database (NIOSHTIC-2) of occupational safety and health publications, documents, grant reports, and other communication products.

PROFESSIONAL CREDENTIALS

Australia

In 2005, the Australian Institute of Occupational Hygiene (AIOH) has accredited professional occupational hygienist through a certification scheme. Occupational Hygienists in Australian certified through this scheme are entitled to use the phrase Certified Occupational Hygienist (COH) as part of their qualifications.

United States of America

Practitioners who successfully meet specific education and work-experience requirements, and pass a written examination administered by the American Board of Industrial Hygiene (ABIH) are authorized to use the term Certified Industrial Hygienist (CIH) or Certified Associate Industrial Hygienist (CAIH). Both of these terms have been codified into law in many states in the United States to identify minimum qualifications of individuals having oversight over certain activities that may affect employee and general public health.

After the initial certification, the CIH or CAIH maintains their certification by meeting on-going requirements for ethical behavior, education, and professional activities (e.g., active practice, technical committees, publishing, teaching).

ABIH certification examinations are offered during a spring and fall testing window each year at more than 400 locations worldwide.

The CIH designation is the most well known and recognized industrial hygiene designation throughout the world. There are approximately 6600 CIHs in the world making ABIH the largest industrial hygiene certification firm. The CAIH certification program was discontinued in 2006. Those who were certified as a CAIH retain their certification through ongoing certification maintenance. People who are currently certified by the ABIH can be found in a public roster.

The ABIH is a recognized certification board by the International Occupational Hygiene Association (IOHA). The CIH certification has been accredited internationally by the International Organization for Standardization/International Electrotechnical Commission (ISO/IEC 17024) (see ANSI). In the United States, the CIH has been accredited by the Council of Engineering and Scientific Specialty Boards [CESB].

The Association of Professional Industrial Hygienists, Inc. (APIH) was established in 1994 to offer credentialing to industrial hygienists who meet the education and experience requirements found in Tennessee Code Annotated, Title 62 APIH adopted the Tennessee Code as its basis for credentialing because it was the first legal definition in the United States of an industrial hygienist in terms of education and experience. The APIH Registration Committee investigates and verifies, through electronic means or correspondence, both educational and experience accomplishments claimed by each applicant for registration. The Committee determines the appropriate level of registration, Registered Industrial

Hygienist or Registered Professional Industrial Hygienist, and then authorizes the registration certificate to be issued.

Canada

In Canada, a practitioner who successfully completes a written and an interview administered by the Canadian Registration Board of Occupational Hygienists can be recognized as a Registered Occupational Hygienist (ROH) or Registered Occupational Hygiene Technician (ROHT). There is also designation to be recognized as a Canadian Registered Safety Professional (CRSP).

United Kingdom

The Faculty of Occupational Hygiene, part of the British Occupational Hygiene Society, represents the interests of professional occupational hygienists.

Membership of the Faculty of Occupational Hygiene is confined to BOHS members who hold a recognized professional qualification in occupational hygiene.

There are three grades of Faculty membership:

- Licentiate (LFOH) holders will have obtained the BOHS Certificate of Operational Competence in Occupational Hygiene and have at least three years' practical experience in the field.
- Members (MFOH) are normally holders of the Diploma of Professional Competence in Occupational Hygiene and have at least five years' experience at a senior level.
- Fellows (FFOH) are senior members of the profession who have made a distinct contribution to the advancement of occupational hygiene.
- All Faculty members participate in a Continuous Professional Development (CPD) scheme designed to maintain a high level of current awareness and knowledge in occupational hygiene.

India

Indian Society of Industrial hygiene was formed in 1981 at Chennai India. Subsequently its secretariat was shifted to Kanpur. The society has registered about 400 members, about 90 of whom life members. The society publishes a newsletter "Industrial Hygiene Link". The current address of the secretary of the society is Shyam Singh Gautam, Secretary, Indian Society of Industrial Hygiene, 11, Shakti Nagar, Rama Devi, Kanpur 2008005 Mobile number 8005187037.

EXAMPLES OF OCCUPATIONAL HYGIENE

- Analysis of physical hazards such as noise that would lead to use of earplugs or earmuffs.

- Planning procedures to protect against infectious disease exposure in the event of a flu pandemic.
- Monitoring and testing the air for hazardous contaminants that can lead to potential worker illness and/or death.

Examples of occupational hygiene careers

- Compliance officer on behalf of regulatory agency
- Professional working on behalf of company for the protection of the workforce
- Consultant working on behalf of companies
- Researcher performing laboratory or field occupational hygiene work

Standard References

It is difficult to create a comprehensive list of references for Occupational Hygiene. Firstly, a list is required for the practices and methodologies involved with the profession of Occupational Hygiene. This list alone can be quite extensive. Secondly, a list of references for each subject area, issue or problem to be resolved from an Occupational Hygiene stand point is required and this information is to be applied to each workplace within each regulatory framework. More importantly the list of references will change due to both changes in technology and changes in requirements for Regulatory compliance. The reference list below may help to get started in self-educating, researching a problem or resolving an issue.

CASE STUDY OF BERKELEY LAB

Roles and Responsibilities

Role	Responsibility
Potentially Exposed Workers	• Request an Exposure Assessment if there is concern about potential exposure, and follow all guidance provided in training and Work Processes to evaluate and control exposures
Supervisors and Work Leads	• Request an Exposure Assessment if there is concern about potential exposure • Ensure that persons within their areas of responsibility comply with this policy and its implementing documents, and have completed the required training prior to beginning work
Industrial Hygiene Subject Matter Expert	• Is responsible for development, approval, revision, and administration of this policy and its implementing documents
Industrial Hygienist	• Facilitates appropriate Exposure Assessment

Line Managers	• Ensure that persons within their areas of responsibility comply with this policy and its implementing documents, and notify the Industrial Hygiene Group of process changes that may affect employee exposures

Definitions

Term	Definition
Activity	For purposes of workplace evaluations and setting priorities, a job (or portion of a job) involving a discrete agent or set of agents to which workers may be exposed. The word "task" is sometimes used in a similar manner.
Baseline Exposure Assessment	A Baseline Exposure Assessment is a process to screen activities to help determine associated risks and hazards. These assessments are generally qualitative, although some quantitative data (collection or review) may be involved.
Chemical Agents	Include all chemicals used at the Laboratory (or in Laboratory-sponsored work). This includes pure chemicals; mixtures (such as paint or cleaning agents); and materials such as asbestos, silica, and engineered nanomaterials.
Engineered Nanomaterials	Discrete materials having structures with at least one dimension between 1 and 100 nanometers, and intentionally created, as opposed to those that are naturally or incidentally formed. Engineered nanomaterials do not include larger materials that may have nanoscale features (*e.g.*, etched silicon wafers), biomolecules (*e.g.*, proteins, nucleic acids, and carbohydrates), or materials with Occupational Exposure Limits (OELs) that address nano-size particles for that substance.
Exposure	Inhalation, ingestion, absorption, injection, or contact with a chemical, biological, or physical agent
Exposure Assessment	The process of defining exposure profiles and judging the acceptability of workplace exposures to environmental agents. These assessments may be quantitative, semiquantitative, or qualitative. These assessments are generally conducted by an Environment, Health & Safety (EHS) professional, which may include industrial hygienists or safety engineers. These assessments may be conducted for representative employees and are not required for each individual. In all cases, employees have full access to exposure-monitoring information, including situations where an individual's exposure is not monitored.
Hazard Assessment	A preliminary evaluation (or screening) of an activity to determine if a more comprehensive Exposure Assessment is required. Hazard Assessments can be performed by work leads, supervisors, workers, or an EHS professional. Hazard Assessments are one form of Baseline Exposure Assessment.

Industrial Hygiene	The art and science of anticipation, recognition, evaluation, and control of occupational health hazards (including exposures to chemicals, noise, and non-ionizing radiation)
Occupational Exposure Limit (OEL)	The maximum concentration of an air contaminant to which working people can be exposed for a specified time interval, usually the maximum average exposure allowed throughout an entire eight-hour shift. OELs are typically Permissible Exposure Limits (PELs) or Threshold Limit Values (TLVs), which are defined in this section. In the absence of formally recognized or regulatory-defined OELs, a chemical manufacturer may establish an appropriate exposure limit. Alternatively, the occupational health staff will determine or develop an appropriate protective level. This process often involves industrial hygiene, occupational medicine, and toxicology staff members. The National Institute for Occupational Safety and Health also publishes Recommended Exposure Limits (RELs), which may be evaluated for use.
Permissible Exposure Limit (PEL)	The OSHA Permissible Exposure Limits are exposure levels considered safe for employee exposure in the workplace. Permissible exposure limits for airborne concentrations of hazardous materials are listed in 29 CFR 1910, Subpart Z, and 29 CFR 1926, Subpart Z; and for physical agents (*e.g.*, noise and non-ionizing radiation) in 29 CFR 1910, Subpart G.
Physical Agents	Agents such as noise, hot and cold extremes, and non-ionizing radiation (*e.g.*, radio-frequency, electromagnetic, microwave, and magnetic fields). Laser exposure is addressed by the Laser Safety Program (See the ES&H Manual *Laser Safety program*).
Professional Judgment	The application and appropriate use of knowledge gained from formal education, experience, observation, experimentation, inference, peer review, and analogy. It allows an experienced industrial hygienist with incomplete or a minimum amount of data to estimate worker exposure in nearly any scenario (adapted from DOE Guide G 440.1 and AIHA's A Strategy for Assessing and Managing Occupational Exposures, third edition), although such judgments and their basis should be documented.
Qualitative Exposure Assessment	The estimation of exposure determinants based on integration of available information and professional judgment (adapted from DOE Guide G 440.1-3, Occupational Exposure Assessment)
Quantitative Exposure Assessment	The determination of exposure based on collection and quantitative analysis of data sufficient to adequately characterize exposures (adapted from DOE Guide G 440.1-3,Occupational Exposure Assessment, and AIHA's A Strategy for Assessing and Managing Occupational Exposures, third edition)
Threshold Limit Values (TLVs)	Airborne concentrations of materials to which nearly all workers may be repeatedly exposed without adverse effect. These values are developed and published by the American Conference of Governmental Industrial Hygienists (ACGIH). Different values are established for eight-hour time-weighted averages, ceilings, and Short-Term Exposure Limits (STELs). Other TLVs are available for nonchemical exposures, such as noise and non-ionizing radiation.

Required Work Processes

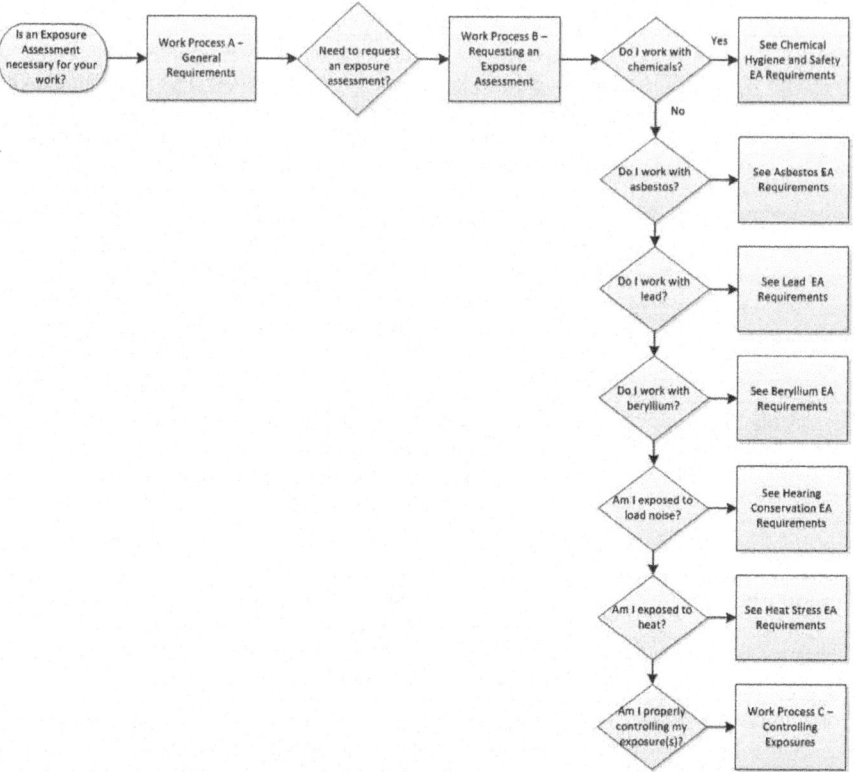

Work Process : A. General Requirements

Every Berkeley Lab worker has a right to a safe workplace. In addition to physical hazards, this includes protection from overexposure to chemicals. Laboratory requirements for safe exposure to chemicals are set forth in 10 CFR 851, DOE's Worker Safety and Health Program rule. This regulation establishes safe work practices, necessary training, and allowable limits to help assure worker protection. One part of the requirements (10 CFR 851.21) specifies the need to identify and assess workplace hazards, including the potential for employee exposure. These assessments can be conducted using multiple methods, including representative air monitoring, estimation, or other modeling.

Based on standard practices, some routine activities, such as use of small quantities of low-hazard chemicals in enclosed systems, may not require additional evaluation. Making these determinations is referred to as Hazard Assessment and is initiated and completed as part of the Job Hazards Analysis (JHA). A more rigorous process is Exposure Assessment, which involves a systematic evaluation of a work activity. These assessments are normally conducted by Berkeley Lab industrial hygienists from the EHS Division and may be initiated from the Work Planning and Control Process (such as JHAs), employee requests, walkthroughs, or other means. Some activities have been characterized and may not need further evaluation.

Berkeley Lab's Industrial Hygiene Group is updating the Exposure Assessment Program and the JHA to assure that potential exposures are evaluated and that workers are properly protected. For more information on Exposure Assessment, contact your division safety coordinator or EHS Liaison, or refer to the Berkeley Lab Exposure Assessment Program Web site. Exposure Assessments may be Qualitative or Quantitative:

- Qualitative Baseline Exposure Assessments are often performed to determine whether a more in-depth Quantitative Exposure Assessment is necessary.
- Quantitative Exposure Assessments include air monitoring, noise dosimetry, and magnetic surveys. They are performed when it is impossible to determine whether a safe level of exposure may be maintained.

The Exposure Assessment Program should be evaluated for effectiveness. This is routinely done through the Laboratory's ESH Technical Assurance Program.

Work Process : B. Requesting an Exposure Assessment

Berkeley Lab workers have a right to records that may be related to their exposure or medical status. This includes Hazard and Exposure Assessments, results of medical surveillance, and other exposure-related information, such as Material Safety Data Sheets.

Work Process : C. Controlling Exposures

Prior to handling chemicals or other materials that might produce harmful exposures, or in areas where there are loud noises or temperature extremes, workers must complete training that covers (a) hazards and (b) methods for controlling the hazards. Examples of such training courses include: EHS0345, Chemical Hygiene for Facilities; EHS0348, Chemical Hygiene and Safety; EHS0356, Nano Safety for Crafts and Technical Work; EHS0310, Respirator Training; and EHS0330, Lead Worker Training. Users of hazardous chemicals/materials must follow training guidance and written procedures covering:

1. Exposure controls
2. Use of controls for chemical handling, including PPE

Controls should be protective at least to the level of Permissible Exposure Limits (PELs) and Threshold Limit Values (TLVs). Controls may include one or a combination of the following:

1. Hazard Elimination, such as:
   a. Eliminating the agent
   b. Substituting the agent for one of less or no hazard

2. Engineering Controls, such as:
   a. Shielding
   b. Ventilation
3. Administrative Controls, such as warning signs and postings
4. PPE, such as:
   a. Respiratory protection
   b. Protective gloves

GOVERNMENT REGULATIONS

Laws and regulations are major tools for protecting people and the environment. Congress is responsible for passing laws that govern the United States. To put these laws into effect, Congress authorizes certain government organizations, including the Environmental Protection Agency (EPA), Occupational Safety and Health Administration (OSHA), and the Department of Homeland Security (DHS) to create and enforce regulations.

How are OSHA and Industrial Hygiene Related?

Under the Act, OSHA develops and sets mandatory occupational safety and health requirements applicable to the more than 6 million workplaces in the U.S. OSHA relies on, among many others, industrial hygienists to evaluate jobs for potential health hazards. Developing and setting mandatory occupational safety and health standards involves determining the extent of employee exposure to hazards and deciding what is needed to control these hazards, thereby protecting the workers. Industrial hygienists, or IHs, are trained to anticipate, recognize, evaluate, and recommend controls for environmental and physical hazards that can affect the health and well-being of workers. More than 40 percent of the OSHA compliance officers who inspect America's workplaces are industrial hygienists. Industrial hygienists also play a major role in developing and issuing OSHA standards to protect workers from health hazards associated with toxic chemicals, biological hazards, and harmful physical agents. They also provide technical assistance and support to the agency's national and regional offices. OSHA also employs industrial hygienists who assist in setting up field enforcement procedures, and who issue technical interpretations of OSHA regulations and standards. Industrial hygienists analyze, identify, and measure workplace hazards or stressors that can cause sickness, impaired health, or significant discomfort in workers through chemical, physical, ergonomic, or biological exposures. Two roles of the OSHA industrial hygienist are to spot those conditions and help eliminate or control them through appropriate measures.

What is a Worksite Analysis?

A worksite analysis is an essential first step that helps an industrial hygienist determine what jobs and work stations are the sources of potential problems.

During the worksite analysis, the industrial hygienist measures and identifies exposures, problem tasks, and risks. The most effective worksite analyses include all jobs, operations, and work activities. The industrial hygienist inspects, researches, or analyzes how the particular chemicals or physical hazards at that worksite affect worker health. If a situation hazardous to health is discovered, the industrial hygienist recommends the appropriate corrective actions.

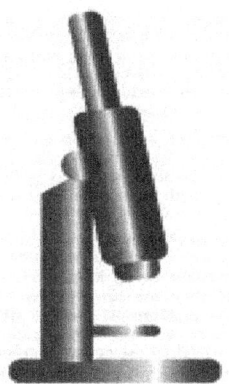

How do IH's Recognize and Control Hazards?

Industrial hygienists recognize that engineering, work practice, and administrative controls are the primary means of reducing employee exposure to occupational hazards. Engineering controls minimize employee exposure by either reducing or removing the hazard at the source or is the worker from the hazard.

Engineering controls include eliminating toxic chemicals and replacing harmful toxic materials with less hazardous ones, enclosing work processes or confining work operations, and installing general and local ventilation systems.

Work practice controls alter the manner in which a task is performed. Some fundamental and easily implemented work practice controls include (1) following proper procedures that minimize exposures while operating production and control equipment; (2) inspecting and maintaining process and control equipment on a regular basis; (3) implementing good house-keeping procedures; (4) providing good supervision and (5) mandating that eating, drinking, smoking, chewing tobacco or gum, and applying cosmetics in regulated areas be prohibited.

Administrative controls include controlling employees' exposure by scheduling production and workers' tasks, or both, in ways that minimize exposure levels. For example, the employer might schedule operations with the highest exposure potential during periods when the fewest employees are present.

When effective work practices and/or engineering controls are not feasible to achieve the permissible exposure limit, or while such controls are being instituted, and in emergencies, appropriate respiratory equipment must be used. In addition, personal protective equipment such as gloves, safety goggles, helmets, safety shoes, and protective clothing may also be required. To be effective, personal protective equipment must be individually selected, properly fitted and periodically refitted; conscientiously and properly worn; regularly maintained; and replaced as necessary.

What Are Some Examples of Job Hazards?

To be effective in recognizing and evaluating on-the-job hazards and recommending controls, industrial hygienists must be familiar with the hazards' characteristics. Major job risks can include air contaminants, and chemical, biological, physical, and ergonomic hazards.

Air Contaminants

These are commonly classified as either particulate or gas and vapor contaminants. The most common particulate contaminants include dusts, fumes, mists, aerosols, and fibers. Dusts are solid particles that are formed or generated from solid organic or inorganic materials by reducing their size through mechanical processes such as crushing, grinding, drilling, abrading or blasting.

Fumes are formed when material from a volatilized solid condenses in cool air. In most cases, the solid particles resulting from the condensation react with air to form an oxide.

The term mist is applied to a finely divided liquid suspended in the atmosphere. Mists are generated by liquids condensing from a vapor back to a liquid or by breaking up a liquid into a dispersed state such as by splashing, foaming or atomizing. Aerosols are also a form of a mist characterized by highly respirable, minute liquid particles.

Fibers are solid particles whose length is several times greater than their diameter.

Gases are formless fluids that expand to occupy the space or enclosure in which they are confined. Examples are welding gases such as acetylene, nitrogen, helium, and argon; and carbon monoxide generated from the operation of internal combustion engines or by its use as a reducing gas in a heat treating operation. Another example is hydrogen sulfide which is formed wherever there is decomposition of materials containing sulfur under reducing conditions.

Liquids change into vapors and mix with the surrounding atmosphere through evaporation. Vapors are the volatile form of substances that are normally in a solid or liquid state at room temperature and pressure. Vapors are the gaseous form of substances which are normally in the solid or liquid state at room temperature and pressure. They are formed by evaporation from a liquid or solid and can be found where parts cleaning and painting takes place and where solvents are used.

Chemical Hazards

Harmful chemical compounds in the form of solids, liquids, gases, mists, dusts, fumes, and vapors exert toxic effects by inhalation (breathing), absorption (through direct contact with the skin), or ingestion (eating or drinking). Airborne chemical hazards exist as concentrations of mists, vapors, gases, fumes, or solids. Some are toxic through inhalation and some of them irritate the skin on contact; some can be toxic by absorption through the skin or through ingestion, and some are corrosive to living tissue.

The degree of worker risk from exposure to any given substance depends on the nature and potency of the toxic effects and the magnitude and duration of exposure.

Information on the risk to workers from chemical hazards can be obtained from the Material Safety Data Sheet (MSDS) that OSHA'S Hazard Communication Standardrequires be supplied by the manufacturer or importer to the purchaser of all hazardous materials. The MSDS is a summary of the important health, safety, and toxicological information on the chemical or the mixture's ingredients. Other provisions of the Hazard Communication Standard require that all containers of hazardous substances in the workplace have appropriate warning and identification labels.

Biological Hazards

These include bacteria, viruses, fungi, and other living organisms that can cause acute and chronic infections by entering the body either directly or through breaks in the skin. Occupations that deal with plants or animals or their products or with food and food processing may expose workers to biological hazards. Laboratory and medical personnel also can be exposed to biological hazards. Any occupations that result in contact with bodily fluids pose a risk to workers from biological hazards.

In occupations where animals are involved, biological hazards are dealt with by preventing and controlling diseases in the animal population as well as proper care and handling of infected animals. Also, effective personal hygiene, particularly proper attention to minor cuts and scratches, especially those on the hands and forearms, helps keep worker risks to a minimum.

In occupations where there is potential exposure to biological hazards, workers should practice proper personal hygiene, particularly hand washing. Hospitals should provide proper ventilation, proper personal protective equipment such as gloves and respirators, adequate infectious waste disposal systems, and appropriate controls including isolation in instances of particularly contagious diseases such as tuberculosis.

Physical Hazards

These include excessive levels of ionizing and nonionizing electromagnetic radiation, noise, vibration, illumination, and temperature.

In occupations where there is exposure to ionizing radiation, time, distance, and shielding are important tools in ensuring worker safety. Danger from radiation increases with the amount of time one is exposed to it; hence, the shorter the time of exposure the smaller the radiation danger.

Distance also is a valuable tool in controlling exposure to both ionizing and non-ionizing radiation. Radiation levels from some sources can be estimated by comparing the squares of the distances between the worker and the source. For example, at a reference point of 10 feet from a source, the radiation is 1/100 of the intensity at 1 foot from the source.

Shielding also is a way to protect against radiation. The greater the protective mass between a radioactive source and the worker, the lower the radiation exposure.

Nonionizing radiation also is dealt with by shielding workers from the source. Sometimes limiting exposure times to nonionizing radiation or increasing the distance is not effective. Laser radiation, for example, cannot be controlled effectively by imposing time limits. An exposure can be hazardous that is faster than the blinking of an eye. Increasing the distance from a laser source may require miles before the energy level reaches a point where the exposure would not be harmful.

Noise, another significant physical hazard, can be controlled by various measures. Noise can be reduced by installing equipment and systems that have been engineered, designed, and built to operate quietly; by enclosing or shielding noisy equipment; by making certain that equipment is in good repair and properly maintained with all worn or unbalanced parts replaced; by mounting noisy equipment on special mounts to reduce vibration; and by installing silencers, mufflers, or baffles.

Substituting quiet work methods for noisy ones is another significant way to reduce noise, for example, welding parts rather than riveting them. Also, treating floors, ceilings, and walls with acoustical material can reduce reflected or reverberant noise. In addition, erecting sound barriers at adjacent work stations around noisy operations will reduce worker exposure to noise generated at adjacent work stations.

t is also possible to reduce noise exposure by increasing the distance between the source and the receiver, by isolating workers in acoustical booths, limiting workers' exposure time to noise, and by providing hearing protection. OSHA requires that workers in noisy surroundings be periodically tested as a precaution against hearing loss.

Another physical hazard, radiant heat exposure in factories such as steel mills, can be controlled by installing reflective shields and by providing protective clothing.

Ergonomic Hazards

The science of ergonomics studies and evaluates a full range of tasks including, but not limited to, lifting, holding, pushing, walking, and reaching. Many ergonomic problems result from technological changes such as increased assembly line speeds, adding specialized tasks, and increased repetition; some problems arise from poorly designed job tasks. Any of those conditions can cause ergonomic hazards such as excessive vibration and noise, eye strain, repetitive motion, and heavy lifting problems. Improperly designed tools or work areas also can be ergonomic hazards. Repetitive motions or repeated shocks over prolonged periods of time as in jobs involving sorting, assembling, and data entry can often cause irritation and inflammation of the tendon sheath of the hands and arms, a condition known as carpal tunnel syndrome.

Ergonomic hazards are avoided primarily by the effective design of a job or jobsite and better designed tools or equipment that meet workers' needs in terms of physical environment and job tasks. Through thorough worksite analyses, employers can set up procedures to correct or control ergonomic hazards by using the appropriate engineering controls (*e.g.*, designing or re-designing work stations, lighting, tools, and equipment); teaching correct work practices (*e.g.*, proper lifting methods); employing proper administrative controls (*e.g.*, shifting workers among several different tasks, reducing production demand, and increasing rest breaks); and, if necessary, providing and mandating personal protective equipment. Evaluating working conditions from an ergonomics standpoint involves looking at the total physiological and psychological demands of the job on the worker.

Overall, industrial hygienists point out that the benefits of a well-designed, ergonomic work environment can include increased efficiency, fewer accidents, lower operating costs. and more effective use of personnel.

In sum, industrial hygiene encompasses a broad spectrum of the working environment. Early in its history OSHA recognized industrial hygiene as an integral part of a healthful work setting. OSHA places a high priority on using industrial hygiene concepts in its health standards and as a tool for effective enforcement of job safety and health regulations. By recognizing and applying the principles of industrial hygiene to the work environment, America's workplaces will become more healthful and safer.

WHAT HELP CAN OSHA PROVIDE?

Safety and Health Program Management Guidelines

Effective management of worker safety and health protection is a decisive factor in reducing the extent and severity of work-related injuries and illnesses and their related costs. To assist employers and employees in developing effective safety and health programs, OSHA published recommended *Safety and Health Program Management Guidelines (Federal Register* 54(18):3908-3916, January 26, 1989). These voluntary guidelines apply to all places of employment covered by OSHA. The guidelines identify four general elements that are critical to the development of a successful safety and health management program:

- management commitment and employee involvement,
- worksite analysis,

- hazard prevention and control, and
- safety and health training.

The guidelines recommend specific actions under each of these general elements to achieve an effective safety and health program. A single free copy of the guidelines can be obtained from the U.S. Department of Labor OSHA/OICA Publications. P.O. Box 37535, Washington, DC 20013-7535, by sending a self-addressed mailing label with your request.

State Programs

The Occupational Safety and Health Act of 1970 encourages states to develop and operate their own job safety and health plans. States administering occupational safety and health programs through plans approved under section 18(b) of the Act, must adopt standards and enforce requirements that are "at least as effective" as federal requirements. There are currently 25 state plan states: 23 cover the private and public sectors (state and local governments) and 2 cover the public sector only. For more information on State Plan states, see the list of states with approved plans at the end of this publication.

Free Onsite Consultation Consultation assistance is available on request to employers who want help in establishing and maintaining a safe and healthful workplace. Largely funded by OSHA, the service is provided at no cost to the employer. Primarily developed for smaller employers with more hazardous operations, the consultation service is delivered by state government agencies or universities employing professional safety consultants and health consultants. Comprehensive assistance includes an appraisal of all work practices and environmental hazards of the workplace and all aspects of the employer's present job safety and health program.

The program is separate from OSHA'S inspection efforts. No penalties are proposed or citations issued for any safety or health problems identified by the consultant. The service is confidential.

For more information concerning consultation assistance, see the list of consultation projects at the end of this publication.

Voluntary Protection Program (VPPs)

Voluntary Protection Programs (VPPs) and onsite consultation services, when coupled with an effective enforcement program, expand worker protection to help meet the goals of the *Act*. The three VPPs--Star, Merit, and Demonstration--are designed to recognize outstanding achievement by companies that have successfully incorporated comprehensive safety and health programs into their total management system. They motivate others to achieve excellent safety and health results in the same outstanding way as they establish a cooperative relationship among employers, employees, and OSHA.

For additional information on VPPs and how to apply, contact the OSHA area or regional offices listed at the end of this publication.

Training and Education

OSHA area offices offer a variety of information services, such as publications, audiovisual aids, technical advice, and speakers for special engagements. The OSHA Training Institute in Des Plaines, IL, provides basic and advanced courses in safety and health for federal and state compliance officers, state consultants, federal agency personnel, and private sector employers, employees, and their representatives.

OSHA also provides funds to nonprofit organizations, through grants to conduct workplace training and education in subjects where OSHA believes there is a lack of workplace training. Grants are awarded annually and grant recipients are expected to contribute 20 percent of the total grant cost. For more information on grants, training, and education, contact the OSHA Training Institute, Office of Training and Education, 1555 Times Drive, Des Plaines, IL 60018; telephone (847) 297-4810.

For further information on any OSHA program, contact your nearest OSHA area or regional office listed at the end of this publication.

Electronic Information

Internet--OSHA standards, interpretations, directives, technical advisors, compliance assistance, and additional information are now on the World Wide Web at http://www.osha.gov/.

CD-ROM--A wide variety of OSHA materials, including standards, interpretations, directives, and more, can be purchased on CD-ROM from the U.S. Government Printing Office. To order, write to the Superintendent of Documents, P.O. Box 371954, Pittsburgh, PA 15250-7954 or telephone (202)512-1800. Specify OSHA Regulations, Documents, and Technical Information on CD-ROM (ORDT), GPO Order No. S/N 729-013-00000-5. The price is $38 per year ($47.50 foreign); $15 per single copy ($18.75 foreign).

CODES OF PRACTICE

Work Health and Safety (WHS) Codes of Practice provide practical guidance to achieve the standards of health, safety and welfare required in the *Work Health and Safety Act 2011* (WHS Act) and work health and safety regulation 2011 (WHS Regulations).

A code of practice applies to anyone who has a duty of care in the circumstances described in the code. In most cases, following an approved code of practice would achieve compliance with the health and safety duties in the WHS Act and Regulations. WHS Codes of Practice are admissible in court proceedings. Courts may regard a code of practice as evidence of what is known about a hazard, risk or control, and rely on it to determine what is reasonably practicable in the circumstances. Compliance with the WHS Act and Regulations may be achieved by following another method, such as a technical or an industry standard, if it provides an equivalent or higher standard of work health and safety than the code.

An inspector may refer to an approved code of practice when issuing an improvement or prohibition notice.

Codes of Practice in Effect in the Commonwealth Jurisdiction

Significant consultation was undertaken during the development of the model WHS Codes of Practice. The following Codes of Practice came into effect in the Commonwealth in 2012:

- How to Manage Work Health and Safety Risks
- Managing the Work Environment and Facilities
- Work Health and Safety Consultation, Co-operation and Co-ordination
- Managing Noise and Preventing Hearing Loss
- Hazardous Manual Tasks
- Confined Spaces
- Managing the Risk of Falls at Workplaces
- Preparation of Safety Data Sheets for Hazardous Chemicals
- Labelling of Workplace Hazardous Chemicals
- How to Manage and Control Asbestos in the Workplace
- How to Safely Remove Asbestos
- First Aid in the Workplace
- Construction Work
- Preventing Falls in Housing Construction
- Managing Electrical Risks at the Workplace
- Managing Risks of Hazardous Chemicals in the Workplace
- Managing Risks of Plant in the Workplace
- Welding Processes
- Excavation Work
- Demolition Work
- Safe Design of Structures
- Spray Painting and Powder Coating
- Abrasive Blasting

Draft Model Codes of Practice

In addition to the Codes of Practice listed above, further Codes of Practice may be introduced into the Commonwealth legislative framework in 2013/14 and will replace any corresponding parts in the Occupational Health and Safety Approved Codes of Practice 2008 (OHS Codes). Until these additional draft model codes are finalised, any applicable remaining OHS Codes will continue to apply in the Commonwealth jurisdiction.

The following draft model Codes of Practice are currently being considered for introduction. They are provided on the Safe Work Australia website as guidance material until approved:

- Working in the vicinity of overhead and underground electric lines
- Safe design, manufacture, import and supply of plant
- Amusement devices
- Scaffolds and scaffolding work
- Tree trimming and removal work - crane access method)
- Industrial lift trucks
- Formwork and falsework
- Managing risks of plant in rural workplaces
- Cranes
- Managing risks in forestry operation
- Managing Cash-in-transit Security Risks
- Traffic Management in Workplaces

Industrial Hygiene: Identification

Industrial hygiene has been defined as "that science and art devoted to the anticipation, recognition, evaluation, and control of those environmental factors or stresses arising in or from the workplace, which may cause sickness, impaired health and well-being, or significant discomfort among workers or among the citizens of the community."

Industrial hygienists use environmental monitoring and analytical methods to detect the extent of worker exposure and employ engineering, work practice controls, and other methods to control potential health hazards.

There has been an awareness of industrial hygiene since antiquity. The environment and its relation to worker health was recognized as early as the fourth century BC when Hippocrates noted lead toxicity in the mining industry. In the first century AD, Pliny the Elder, a Roman scholar, perceived health risks to those working with zinc and sulfur. He devised a face mask made from an animal bladder to protect workers from exposure to dust and lead fumes. In the second century AD, the Greek physician, Galen, accurately described the pathology of lead poisoning and also recognized the hazardous exposures of copper miners to acid mists.

In the Middle Ages, guilds worked at assisting sick workers and their families. In 1556, the German scholar, Agricola, advanced the science of industrial hygiene even further when, in his book *De Re Metallica*, he described the diseases of miners and prescribed preventive measures. The book included suggestions for mine ventilation and worker protection, discussed mining accidents, and described diseases associated with mining occupations such as silicosis.

Industrial hygiene gained further respectability in 1700 when Bernardo Ramazzini, known as the "father of industrial medicine," published in Italy the first comprehensive book on industrial medicine, *De Morbis Artificum Diatriba (The Diseases of Workmen)*. The book contained accurate descriptions of the occupational diseases of most of the workers of his time. Ramazzini greatly affected the future of industrial hygiene because he asserted that occupational diseases should be studied in the work environment rather than in hospital wards.

Industrial hygiene received another major boost in 1743 when Ulrich Ellenborg published a pamphlet on occupational diseases and injuries among gold miners. Ellenborg also wrote about the toxicity of carbon monoxide, mercury, lead, and nitric acid.

In England in the 18th century, Percival Pott, as a result of his findings on the insidious effects of soot on chimney sweepers, was a major force in getting the British Parliament to pass the *Chimney-Sweepers Act of 1788*. The passage of the English Factory Acts beginning in 1833 marked the first effective legislative acts in the field of industrial safety. The Acts, however, were intended to provide compensation for accidents rather than to control their causes. Later, various other European nations developed workers' compensation acts, which stimulated the adoption of increased factory safety precautions and the establishment of medical services within industrial plants.

In the early 20th century in the U.S., Dr. Alice Hamilton led efforts to improve industrial hygiene. She observed industrial conditions first hand and startled mine owners, factory managers, and state officials with evidence that there was a correlation between worker illness and exposure to toxins. She also presented definitive proposals for eliminating unhealthful working condition.

At about the same time, U.S. federal and state agencies began investigating health conditions in industry. In 1908, public awareness of occupationally related diseases stimulated the passage of compensation acts for certain civil employees. States passed the first workers' compensation laws in 1911. And in 1913, the New York Department of Labor and the Ohio Department of Health established the first state industrial hygiene programs. All states enacted such legislation by 1948. In most states, there is some compensation coverage for workers contracting occupational diseases.

The U.S. Congress has passed three landmark pieces of legislation related to safeguarding workers' health: (1) the *Metal and Nonmetallic Mines Safety Act of 1966*, (2) the *Federal Coal Mine Safety and Health Act of 1969*, and (3) the *Occupational Safety and Health Act of 1970 (OSH Act)*. Today, nearly every employer is required to implement the elements of an industrial hygiene and safety, occupational health, or hazard communication program and to be responsive to the Occupational Safety and Health Administration (OSHA) and its regulations.

Osha and Industrial Hygiene

Under the OSH Act, OSHA develops and sets mandatory occupational safety and health requirements applicable to the more than 6 million workplaces in the U.S. OSHA relies on, among many others, industrial hygienists to evaluate jobs for potential health hazards. Developing and setting mandatory occupational safety and health standards involves determining the extent of employee exposure to hazards and deciding what is needed to control these hazards to protect workers. Industrial hygienists are trained to anticipate, recognize, evaluate, and recommend controls for environmental and physical hazards that can affect the health and well-being of workers.

More than 40 percent of the OSHA compliance officers who inspect America's workplaces are industrial hygienists. Industrial hygienists also play a major role in developing and issuing OSHA standards to protect workers from health hazards associated with toxic chemicals, biological hazards, and harmful physical agents. They also provide technical assistance and support to the agency's national and regional offices. OSHA also employs industrial hygienists who assist in setting up field enforcement procedures, and who issue technical interpretations of OSHA regulations and standards.

Industrial hygienists analyze, identify, and measure workplace hazards or stresses that can cause sickness, impaired health, or significant discomfort in workers through chemical, physical, ergonomic, or biological exposures. Two roles of the OSHA industrial hygienist are to spot those conditions and help eliminate or control them through appropriate measures.

Worksite Analysis

A worksite analysis is an essential first step that helps an industrial hygienist determine what jobs and work stations are the sources of potential problems. During the worksite analysis, the industrial hygienist measures and identifies exposures, problem tasks, and risks. The most-effective worksite analyses include all jobs, operations, and work activities. The industrial hygienist inspects, researches, or analyzes how the particular chemicals or physical hazards at that worksite affect worker health. If a situation hazardous to health is discovered, the industrial hygienist recommends the appropriate corrective actions.

Recognizing and Controlling Hazards

Industrial hygienists recognize that engineering, work practice, and administrative controls are the primary means of reducing employee exposure to occupational hazards.

Engineering controls minimize employee exposure by either reducing or removing the hazard at the source or isolating the worker from the hazard. Engineering controls include eliminating toxic chemicals and substituting non-toxic chemicals, enclosing work processes or confining work operations, and the installation of general and local ventilation systems.

Work practice controls alter the manner in which a task is performed. Some fundamental and easily implemented work practice controls include (1) changing existing work practices to follow proper procedures that minimize exposures while operating production and control equipment; (2) inspecting and maintaining process and control equipment on a regular basis; (3) implementing good housekeeping procedures; (4) providing good supervision; and (5) mandating that eating, drinking, smoking, chewing tobacco or gum, and applying cosmetics in regulated areas be prohibited.

Administrative controls include controlling employees' exposure by scheduling production and tasks, or both, in ways that minimize exposure levels. For example, the employer might schedule operations with the highest exposure potential during periods when the fewest employees are present.

When effective work practices or engineering controls are not feasible or while such controls are being instituted, appropriatepersonal protective equipment must be used. Examples of personal protective equipment are gloves, safety goggles, helmets, safety shoes, protective clothing, and respirators. To be effective, personal protective equipment must be individually selected, properly fitted and periodically refitted; conscientiously and properly worn; regularly maintained; and replaced, as necessary.

Examples of Job Hazards

To be effective in recognizing and evaluating on-the-job hazards and recommending controls, industrial hygienists must be familiar with the hazards' characteristics. Potential hazards can include air contaminants, and chemical, biological, physical, and ergonomic hazards.

Air Contaminants

These are commonly classified as either particulate or gas and vapor contaminants. The most common particulate contaminants include dusts, fumes, mists, aerosols, and fibers.

Dusts are solid particles generated by handling, crushing, grinding, colliding, exploding, and heating organic or inorganic materials such as rock, ore, metal, coal, wood, and grain

Fumes are formed when material from a volatilized solid condenses in cool air. In most cases, the solid particles resulting from the condensation react with air to form an oxide.

The term mist is applied to liquid suspended in the atmosphere. Mists are generated by liquids condensing from a vapor back to a liquid or by a liquid being dispersed by splashing or atomizing. Aerosols are also a form of a mist characterized by highly respirable, minute liquid particles.

Fibers are solid particles whose length is several times greater than their diameter, such as asbestos.

Gases are formless fluids that expand to occupy the space or enclosure in which they are confined. They are atomic, diatomic, or molecular in nature as opposed to droplets or particles which are made up of millions of atoms or molecules. Through evaporation, liquids change into vapors and mix with the surrounding atmosphere. Vapors are the volatile form of substances that are normally in a solid or liquid state at room temperature and pressure. Vapors are gases in that true vapors are atomic or molecular in nature.

Chemical Hazards

Harmful chemical compounds in the form of solids, liquids, gases, mists, dusts, fumes, and vapors exert toxic effects by inhalation (breathing), absorption (through direct contact with the skin), or ingestion (eating or drinking). Airborne chemical hazards exist as concentrations of mists, vapors, gases, fumes, or solids. Some are toxic through inhalation and some of them irritate the skin on contact; some can be toxic by absorption through the skin or through ingestion, and some are corrosive to living tissue.

The degree of worker risk from exposure to any given substance depends on the nature and potency of the toxic effects and the magnitude and duration of exposure. Information on the risk to workers from chemical hazards can be obtained from the Material Safety Data Sheet (MSDS) that OSHA's Hazard Communication Standard requires be supplied by the manufacturer or importer to the purchaser of all hazardous materials. The MSDS is a summary of the important health, safety, and toxicological information on the chemical or the mixture's ingredients. Other provisions of the Hazard Communication Standard require that all containers of hazardous substances in the workplace have appropriate warning and identification labels.

Biological Hazards

These include bacteria, viruses, fungi, and other living organisms that can cause acute and chronic infections by entering the body either directly or through breaks in the skin.

Occupations that deal with plants or animals or their products or with food and food processing may expose workers to biological hazards. Laboratory and medical personnel also can be exposed to biological hazards. Any occupations that result in contact with bodily fluids pose a risk to workers from biological hazards.

In occupations where animals are involved, biological hazards are dealt with by preventing and controlling diseases in the animal population as well as properly caring for and handling infected animals. Also, effective personal hygiene, particularly proper attention to minor cuts and scratches especially on the hands and forearms, helps keep worker risks to a minimum.

In occupations where there is potential exposure to biological hazards, workers should practice proper personal hygiene, particularly hand washing. Hospitals should provide proper ventilation, proper personal protective equipment such as gloves and respirators, adequate infectious waste disposal systems, and appropriate controls including isolation in instances of particularly contagious diseases such as tuberculosis.

Physical Hazards

These include excessive levels of ionizing and nonionizing electromagnetic radiation, noise, vibration, illumination, and temperature.

In occupations where there is exposure to ionizing radiation, time, distance, and shielding are important tools in ensuring worker safety. Danger from radiation increases with the amount of time one is exposed to it; hence, the shorter the time of exposure the smaller the radiation danger.

Distance also is a valuable tool in controlling exposure to both ionizing and nonionizing radiation. Radiation levels from some sources can be estimated by comparing the squares of the distances between the worker and the source. For example, at a reference point of 10 feet from a source, the radiation is $1/100$ of the intensity at 1 foot from the source.

Shielding also is a way to protect against radiation. The greater the protective mass between a radioactive source and the worker, the lower the radiation exposure.

In some instances, however, limiting exposure to or increasing distance from certain forms of nonionizing radiation, such as lasers, is not effective. For example, an exposure to laser radiation that is faster than the blinking of an eye can be hazardous and would require workers to be miles from the laser source before being adequately protected. Shielding workers from this source can be an effective control method.

Noise, another significant physical hazard, can be controlled by various measures. Noise can be reduced by installing equipment and systems that have been engineered, designed, and built to operate quietly; by enclosing or shielding noisy equipment; by making certain that equipment is in good repair and properly maintained with all worn or unbalanced parts replaced; by mounting noisy equipment on special mounts to reduce vibration; and by installing silencers, mufflers, or baffles.

Substituting quiet work methods for noisy ones is another significant way to reduce noise-for example, welding parts rather than riveting them. Also, treating floors, ceilings, and walls with acoustical material can reduce reflected or reverberant noise. In addition, erecting sound barriers at adjacent work stations around noisy operations will reduce worker exposure to noise generated at adjacent work stations.

It is also possible to reduce noise exposure by increasing the distance between the source and the receiver, by isolating workers in acoustical booths, limiting workers' exposure time to noise, and by providing hearing protection. OSHA requires that workers in noisy surroundings be periodically tested as a precaution against hearing loss.

Another physical hazard, radiant heat exposure in factories such as steel mills, can be controlled by installing reflective shields and by providing protective clothing.

Ergonomic Hazards

The science of ergonomics studies and evaluates a full range of tasks including, but not limited to, lifting, holding, pushing, walking, and reaching. Many ergonomic problems result from technological changes such as increased assembly line speeds, adding specialized tasks, and increased repetition; some problems arise from poorly designed job tasks. Any of those conditions can cause ergonomic hazards such as excessive vibration and noise, eye strain, repetitive motion, and heavy lifting problems. Improperly designed tools or work areas also can be ergonomic hazards. Repetitive motions or repeated shocks over prolonged periods of time as in jobs involving sorting, assembling, and data entry can often cause irritation and inflammation of the tendon sheath of the hands and arms, a condition known as carpal tunnel syndrome.

Ergonomic hazards are avoided primarily by the effective design of a job or jobsite and by better designed tools or equipment that meet workers' needs in terms of physical environment and job tasks. Through thorough worksite analyses, employers can set up procedures to correct or control ergonomic hazards by using the appropriate engineering controls (*e.g.*, designing or redesigning work stations, lighting, tools, and equipment); teaching correct work practices (*e.g.*, proper lifting methods); employing proper administrative controls (*e.g.*, shifting workers among several different tasks, reducing production demand, and increasing rest breaks); and, if necessary, providing and mandating personal protective equipment. Evaluating working conditions from an ergonomics standpoint involves looking at the total physiological and psychological demands of the job on the worker.

Overall, industrial hygienists point out that the benefits of a well-designed, ergonomic work environment can include increased efficiency, fewer accidents, lower operating costs, and more effective use of personnel.

In sum, industrial hygiene encompasses a broad spectrum of the working environment. Early in its history, OSHA recognized industrial hygiene as an integral part of a healthful work setting. OSHA places a high priority on using industrial hygiene concepts in its health standards and as a tool for effective enforcement of job safety and health regulations. By recognizing and applying the principles of industrial hygiene to the work environment, America's workplaces will become more healthful and safer.

Industrial Hygiene Control

After a hazard has been recognized and evaluated, the most appropriate interventions (methods of control) for a particular hazard must be determined. Control methods usually fall into three categories:

1. engineering controls
2. administrative controls
3. personal protective equipment.

As with any change in work processes, training must be provided to ensure the success of the changes.

Engineering controls are changes to the process or equipment that reduce or eliminate exposures to an agent. For example, substituting a less toxic chemical in a process or installing exhaust ventilation to remove vapours generated during a process step, are examples of engineering controls. In the case of noise control, installing sound-absorbing materials, building enclosures and installing mufflers on air exhaust outlets are examples of engineering controls. Another type of engineering control might be changing the process itself. An example of this type of control would be removal of one or more degreasing steps in a process that originally required three degreasing steps. By removing the need for the task that produced the exposure, the overall exposure for the worker has been controlled. The advantage of engineering controls is the relatively small involvement of the worker, who can go about the job in a more controlled environment when, for instance, contaminants are automatically removed from the air. Contrast this to the situation where the selected method of control is a respirator to be worn by the worker while performing the task in an "uncontrolled" workplace. In addition to the employer actively installing engineering controls on existing equipment, new equipment can be purchased that contains the controls or other more effective controls. A combination approach has often been effective (*i.e.*, installing some engineering controls now and requiring personal protective equipment until new equipment arrives with more effective controls that will eliminate the need for personal protective equipment). Some common examples of engineering controls are:

- ventilation (both general and local exhaust ventilation)
- isolation (place a barrier between the worker and the agent)
- substitution (substitute less toxic, less flammable material, *etc.*)
- change the process (eliminate hazardous steps).

The occupational hygienist must be sensitive to the worker's job tasks and must solicit worker participation when designing or selecting engineering controls. Placing barriers in the workplace, for example, could significantly impair a worker's ability to perform the job and may encourage "work arounds". Engineering controls are the most effective methods of reducing exposures. They are also, often, the most expensive. Since engineering controls are effective and expensive it is important to maximize the involvement of the workers in the selection and design of the controls. This should result in a greater likelihood that the controls will reduce exposures.

Administrative controls involve changes in how a worker accomplishes the necessary job tasks — for example, how long they work in an area where exposures occur, or changes in work practices such as improvements in body positioning to reduce exposures. Administrative controls can add to the effectiveness of an intervention but have several drawbacks:

1. Rotation of workers may reduce overall average exposure for the workday but it provides periods of high short-term exposure for a larger number of workers. As more becomes known about toxicants and their modes of action, short-term peak exposures may represent a greater risk than would be calculated based on their contribution to average exposure.

2. Changing work practices of workers can present a significant enforcement and monitoring challenge. How work practices are enforced and monitored determines whether or not they will be effective. This constant management attention is a significant cost of administrative controls.

Personal protective equipment consists of devices provided to the worker and required to be worn while performing certain (or all) job tasks. Examples include respirators, chemical goggles, protective gloves and faceshields. Personal protective equipment is commonly used in cases where engineering controls have not been effective in controlling the exposure to acceptable levels or where engineering controls have not been found to be feasible (for cost or operational reasons). Personal protective equipment can provide significant protection to workers if worn and used correctly. In the case of respiratory protection, protection factors (ratio of concentration outside the respirator to that inside) can be 1,000 or more for positive-pressure supplied air respirators or ten for half-face air-purifying respirators. Gloves (if selected appropriately) can protect hands for hours from solvents. Goggles can provide effective protection from chemical splashes.

Intervention: Factors to Consider

Often a combination of controls is used to reduce the exposures to acceptable levels. Whatever methods are selected, the intervention must reduce the exposure and resulting hazard to an acceptable level. There are, however, many other factors that need to be considered when selecting an intervention. For example:

*   effectiveness of the controls
*   ease of use by the employee
*   cost of the controls
*   adequacy of the warning properties of the material
*   acceptable level of exposure

- frequency of exposure
- route(s) of exposure
- regulatory requirements for specific controls.

Effectiveness of Controls

Effectiveness of controls is obviously a prime consideration when taking action to reduce exposures. When comparing one type of intervention to another, the level of protection required must be appropriate for the challenge; too much control is a waste of resources. Those resources could be used to reduce other exposures or exposures of other employees. On the other hand, too little control leaves the worker exposed to unhealthy conditions. A useful first step is to rank the interventions according to their effectiveness, then use this ranking to evaluate the significance of the other factors.

Ease of Use

For any control to be effective the worker must be able to perform his or her job tasks with the control in place. For example, if the control method selected is substitution, then the worker must know the hazards of the new chemical, be trained in safe handling procedures, understand proper disposal procedures, and so on. If the control is isolation – placing an enclosure around the substance or the worker – the enclosure must allow the worker to do his or her job. If the control measures interfere with the tasks of the job, the worker will be reluctant to use them and may find ways to accomplish the tasks that could result in increased, not decreased, exposures.

Cost

Every organization has limits on resources. The challenge is to maximize the use of those resources. When hazardous exposures are identified and an intervention strategy is being developed, cost must be a factor. The "best buy" many times will not be the lowest- or highest-cost solutions. Cost becomes a factor only after several viable methods of control have been identified. Cost of the controls can then be used to select the controls that will work best in that particular situation. If cost is the determining factor at the outset, poor or ineffective controls may be selected, or controls that interfere with the process in which the employee is working. It would be unwise to select an inexpensive set of controls that interfere with and slow down a manufacturing process. The process then would have a lower throughput and higher cost. In very short time the "real" costs of these "low cost" controls would become enormous. Industrial engineers understand the layout and overall process; production engineers understand the manufacturing steps and processes; the financial analysts understand the resource allocation problems. Occupational hygienists can provide a unique insight into these discussions due to their understanding of the specific employee's job tasks, the employee's interaction with the manufacturing equipment as well as how the controls will work in a particular setting. This team approach increases the likelihood of selecting the most appropriate (from a variety of perspectives) control.

Adequacy of Warning Properties

When protecting a worker against an occupational health hazard, the warning properties of the material, such as odour or irritation, must be considered. For example, if a semiconductor worker is working in an area where arsine gas is used, the extreme toxicity of the gas poses a significant potential hazard. The situation is compounded by arsine's very poor warning properties – the workers cannot detect the arsine gas by sight or smell until it is well above acceptable levels. In this case, controls that are marginally effective at keeping exposures below acceptable levels should not be considered because excursions above acceptable levels cannot be detected by the workers. In this case, engineering controls should be installed to isolate the worker from the material. In addition, a continuous arsine gas monitor should be installed to warn workers of the failure of the engineering controls. In situations involving high toxicity and poor warning properties, preventive occupational hygiene is practised. The occupational hygienist must be flexible and thoughtful when approaching an exposure problem.

Acceptable Level of Exposure

If controls are being considered to protect a worker from a substance such as acetone, where the acceptable level of exposure may be in the range of 800 ppm, controlling to a level of 400 ppm or less may be achieved relatively easily. Contrast the example of acetone control to control of 2-ethoxyethanol, where the acceptable level of exposure may be in the range of 0.5 ppm. To obtain the same per cent reduction (0.5 ppm to 0.25 ppm) would probably require different controls. In fact, at these low levels of exposure, isolation of the material may become the primary means of control. At high levels of exposure, ventilation may provide the necessary reduction. Therefore, the acceptable level determined (by the government, company, *etc.*) for a substance can limit the selection of controls.

Frequency of Exposure

When assessing toxicity the classic model uses the following relationship:

TIME x CONCENTRATION = DOSE

Dose, in this case, is the amount of material being made available for absorption. The previous discussion focused on minimizing (lowering) the concentration portion of this relationship. One might also reduce the time spent being exposed (the underlying reason for administrative controls). This would similarly reduce the dose. The issue here is not the employee spending time in a room, but how often an operation (task) is performed. The distinction is important. In the first example, the exposure is controlled by removing the workers when they are exposed to a selected amount of toxicant; the intervention effort is not directed at controlling the amount of toxicant (in many situations there may be a combination approach).

In the second case, the frequency of the operation is being used to provide the appropriate controls, not to determine a work schedule. For example, if an operation such as degreasing is performed routinely by an employee, the controls may include ventilation, substitution of a less toxic solvent or even automation of the process. If the operation is performed rarely (*e.g.*, once per quarter) personal protective equipment may be an option (depending on many of the factors described in this section). As these two examples illustrate, the frequency with which an operation is performed can directly affect the selection of controls. Whatever the exposure situation, the frequency with which a worker performs the tasks must be considered and factored into the control selection.

Route of exposure obviously is going to affect the method of control. If a respiratory irritant is present, ventilation, respirators, and so on, would be considered. The challenge for the occupational hygienist is identifying all routes of exposure. For example, glycol ethers are used as a carrier solvent in printing operations. Breathing-zone air concentrations can be measured and controls implemented. Glycol ethers, however, are rapidly absorbed through intact skin. The skin represents a significant route of exposure and must be considered. In fact, if the wrong gloves are chosen, the skin exposure may continue long after the air exposures have decreased (due to the employee continuing to use gloves that have experienced breakthrough). The hygienist must evaluate the substance—its physical properties, chemical and toxicological properties, and so on—to determine what routes of exposure are possible and plausible (based on the tasks performed by the employee).

In any discussion of controls, one of the factors that must be considered is the regulatory requirements for controls. There may well be codes of practice, regulations, and so on, that require a specific set of controls. The occupational hygienist has flexibility above and beyond the regulatory requirements, but the minimum mandated controls must be installed. Another aspect of the regulatory requirements is that the mandated controls may not work as well or may conflict with the best judgement of the occupational hygienist. The hygienist must be creative in these situations and find solutions that satisfy the regulatory as well as best practice goals of the organization.

Training and Labelling

Regardless of what form of intervention is eventually selected, training and other forms of notification must be provided to ensure that the workers understand the interventions, why they were selected, what reductions in exposure are expected, and the role of the workers in achieving those reductions. Without the participation and understanding of the workforce, the interventions will likely fail or at least operate at reduced efficiency. Training builds hazard awareness in the workforce. This new awareness can be invaluable to the occupational hygienist in identifying and reducing previously unrecognized exposures or new exposures.

Training, labelling and related activities may be part of a regulatory compliance scheme. It would be prudent to check the local regulations to ensure that whatever type of training or labelling is undertaken satisfies the regulatory as well as operational requirements.

CONCLUSION

In this short discussion on interventions, some general considerations have been presented to stimulate thought. In practice, these rules become very complex and often have significant ramifications for employee and company health. The occupational hygienist's professional judgement is essential in selecting the best controls. Best is a term with many different meanings. The occupational hygienist must become adept at working in teams and soliciting input from the workers, management and technical staff.

Chapter 4

CHEMICAL PROCESS: SOURCE MODELS

INTRODUCTION TO SOURCE MODELS

Most accidents in chemical plants result in spills of toxic, flammable, and explosives materials. Source models are an important part of the consequence modeling procedure shown in More details are provided elsewhere. Accidents begin with an incident, which usually results in the loss of containment of material from the process. The material has hazardous properties, which might include toxic properties and energy content. Typical incidents might include the rupture or break of a pipeline, a hole in a tank or pipe, runaway reaction, or fire external to the vessel. Once the incident is known, source models are selected to describe how materials are discharged from the process. The source model provides a description of the rate of discharge, the total quantity discharged (or total time of discharge), and the state of the discharge (that is, solid, liquid, vapor, or a combination). A dispersion model is subsequently used to describe how the material is transported downwind and dispersed to some concentration levels. For flammable releases fire and explosion models convert the source model information on the release into energy hazard potentials, such as thermal radiation and explosion overpressures.

Effect models convert these incident-specific results into effects on people (injury or death) and structures. Environmental impacts could also be considered, but we do not do so here. Additional refinement is provided by mitigation factors, such as water sprays, foam systems, and sheltering or evacuation, which tend to reduce the magnitude of potential effects in real incidents. Source models are constructed from fundamental or empirical equations representing the physicochemical processes occurring during the release of materials. For a reasonably complex.

Chemical Process Safety Model

December 3, 1984, water contamination of a tank of methyl isocyanate in Bhopal, India initiated a series of events that let to a catastrophic toxic release, killing more than 3,000 residents and injuring over 100,000.

Immediately after, leaders from the chemical industry asked AIChE to lead a collaborative effort to eliminate catastrophic process incidents by advancing state of the art technology and management practices, serving as the premier resource for information on process safety, supporting process safety in engineering, and promoting process safety as a key industry value. And so began the industry response to the tragic event.

On March 25, 1985, AIChE formed the Center for Chemical Process Safety with seventeen charter member companies. CCPS quickly set out to publish its first process safety guideline book, Guidelines for Hazard Evaluation Procedures, and by 1990, more than a dozen guideline books had been published, along with CCPS' call to action publication, A Challenge to Commitment. In these initial publications, CCPS first codified the critical elements of process safety and provided key tools to continually improve process safety programs. Focused workshops and international conferences provided opportunities for communal learning and discussions regarding process safety and for the exploration of new ideas and important developments relevant to the chemical industry.

CCPS continues to address the most important process safety needs and encourage an overall culture of process safety. Over 100 members now participate in CCPS, including most of the world's leading chemical, petroleum, pharmaceutical and related manufacturing companies. CCPS' extensive body of work marks the progress made in these areas and it continues to expand a catalog of over 100 books and products, build on a legacy of 21 successful international conferences, and cultivate the Safety in Chemical Engineering Education (SAChE) university curriculum program.

A Challenge to Commitment ...and reaching forward

The world has changed over the last two decades. In today's global economy, technology that was unheard of in 1984 is now readily available. Little-developed parts of the globe have become major production centers. Specialty chemicals have become commodities, while new applications have arisen. Through all these changes, the job of process safety has evolved, but the need remains constant.

Since 1984 CCPS has helped industry develop tools to keep our workplaces and communities safer, even as technology and businesses have become more complex. The results have been heartening, but also sobering. Since 1984, there has not been another accident having as strong an industry-wide impact or receiving as much global attention as Bhopal. According to the US Occupational Safety and Health Administration, since 1992 on site fatalities from process safety incidents have dropped by over 60%. Fatalities, when they occur, have been contained to the plant site, and when off-site releases have occurred, emergency procedures have kept injuries low.

On the other hand, the US Chemical Safety Board has identified 167 incidents in the last 25 years based on chemical reactivity alone, mostly at smaller manufacturers and companies whose main business was not chemistry. Nearly all of these incidents could have been prevented had basic process safety guidelines and references been consulted.

As CCPS becomes a global organization, disseminating process safety resources and information into China, India and South America, the mission remains clear. We must continue to be a beacon within the chemical, petroleum, pharmaceutical and related industries. To achieve this, we must evolve as an organization in order to maintain relevance and offer utility within a sector that is itself constantly changing. Second, we must expand this central core of companies and work to find common threads to better promote the culture of process safety. Our work is far from done.

PROCESS SAFETY MANAGEMENT

Process safety management is a regulation, promulgated by the U.S. Occupational Safety and Health Administration (OSHA). A process is any activity or combination of activities including any use, storage, manufacturing, handling or the on-site movement of highly hazardous chemicals (HHCs) as defined by OSHA and the Environmental Protection Agency.

Definition

Process safety management is an analytical tool focused on preventing releases of any substance defined as a "highly hazardous chemicals" by the EPA or OSHA. Process Safety Management (PSM) refers to a set of inter-related approaches to manage hazards associated with the process industries and is intended to reduce the frequency and severity of incidents resulting from releases of chemicals and other energy sources (US OSHA 1993). These standards are composed of organizational and operational procedures, design guidance, audit programs, and a host of other methods.

Elements of Process Safety Management (PSM)

The process safety management program is divided into 14 elements. The U.S. Occupational Safety and Health Administration (OSHA) 1910.119 define all 14 elements of process safety management plan.

14 Elements of OSHA Process Safety Management Program (PSM)

- Process Safety Information
- Process Hazard Analysis
- Operating Procedures
- Training
- Contractors
- Mechanical Integrity
- Hot Work
- Management of Change
- Incident Investigation
- Compliance Audits
- Trade Secrets
- Employee Participation
- Pre-startup Safety Review
- Emergency Planning and Response

All of those elements mentioned above are interlinked and interdependent. There is a tremendous interdependency of the various elements of PSM. All elements are related and are necessary to make up the entire PSM picture. Every element either contributes information to other elements for the completion or utilizes information from other elements in order to be completed.

Process Safety Information (PSI)

PSI or process safety information might be considered the keystone of a PSM Program in that it tells you what you are dealing with from both the equipment and the process standpoint. In order to be in compliance with the OSHA PSM regulations the process safety information should include information pertaining to the hazards of the highly hazardous chemicals used or produced by the process, information pertaining to the technology of the process and information pertaining to the equipment in the process.

Information pertaining to the hazards of the highly hazardous chemicals in the process should consist of at least the following:

- Toxicity information
- Permissible exposure limit
- Physical data
- Reactivity data
- Corrosivity data
- Thermal and chemical stability data
- Hazardous effects of inadvertent mixing of different materials that could foreseeably occur.

Information pertaining to the technology of the process should include at least the following:

- A block flow diagram or simplified process flow diagram
- Process chemistry and its properties
- Maximum intended inventory
- Safety upper and lower limits for such items as temperatures, pressures, flows or compositions
- An evaluation of the consequences of deviations, including those effecting the safety and health of the employees.

Information pertaining to the equipment in the process should include following:

- Materials of construction
- Piping and instrument diagram (P&ID's)
- Electrical classification
- Relief system design and design basis
- Ventilation system design
- Design codes and standards employed
- Material and energy balances for processes built after May 26, 1992
- Safety system (for example interlocks, detection or suppression systems)

The employer should document that equipment complies with recognized and generally accepted good engineering practices (RAGAGEP) For existing equipment designed and constructed in accordance with codes, standards or practices that are no longer in general use, the employer should determine and document that the equipment is designed, maintained, inspected, tested and operating in a safe manner.

Compliance

A process includes any group of vessels which are interconnected or separate and contain Highly Hazardous Chemicals (HHC's) which could be involved in a potential release. A process safety incident is the "Unexpected release of toxic, reactive, or flammable liquids and gases in processes involving highly hazardou chemicals. Incidents continue to occur in various industries that use highly hazardous chemicals which exhibit toxic, reactive, flammable, or even explosive properties, or may exhibit a combination of these properties. Regardless of the industry that uses these highly hazardous chemicals, there is a potential for an accidental release any time they are not properly controlled. This, in turn, creates the possibility of disaster. To help assure safe and healthy workplaces, OSHA has issued the Process Safety Management of Highly Hazardous Chemicals regulation (Title 29 of CFR Section 1910.119)[1] which contains requirements for the management of hazards associated with processes using highly hazardous chemicals."

Any facility that stores or uses a defined "highly hazardous chemical" must comply with OSHA's process safety management (PSM) regulations as well as the quite similar United States Environmental Protection Agency (EPA) Risk management program (RMP) regulations (Title 40 CFR Part 68). The EPA has published a model RMP plan for an ammonia refrigeration facility which provides excellent guidance on how to comply with either OSHA's PSM regulations or the EPA's RMP regulations.

The Center for Chemical Process Safety (CCPS) of the American Institute of Chemical Engineers (AIChE) has published a widely used book that explains various methods for identifying hazards in industrial facilities and quantifying their potential severity. Appendix D of the OSHA's PSM regulations endorses the use of the methods explained in that book. AIChE publishes additional guidelines for process safety documentation, implementing process safety management systems, and the Center for Chemical Process Safety publishes an engineering design for process safety.

In Australia, consideration of process safety management is a key consideration for the management of Major Hazard Facilities (MHFs).

CHEMICAL KINETICS

Chemical kinetics, also known as **reaction kinetics**, is the study of rates of chemical processes. Chemical kinetics includes investigations of how different experimental conditions can influence the speed of a chemical reaction and yield information about the reaction's mechanism and transition states, as well as the construction of mathematical models that can describe the characteristics of a chemical reaction. In 1864, Peter Waage and Cato Guldberg pioneered the development of chemical kinetics by formulating the law of mass action, which states that the speed of a chemical reaction is proportional to the quantity of the reacting substances.

Chemical kinetics deals with the experimental determination of reaction rates from which rate laws and rate constants are derived. Relatively simple rate laws exist for zero-order reactions (for which reaction rates are independent of concentration), first-order reactions, and second-order reactions, and can be derived for others. In consecutive reactions, the rate-determining step often determines the kinetics. In consecutive first-order reactions, a steady state approximation can simplify the rate law. The activation energy for a reaction is experimentally determined through the Arrhenius equation and the Eyring equation. The main factors that influence the reaction rate include: the physical state of the reactants, the concentrations of the reactants, the temperature at which the reaction occurs, and whether or not any catalysts are present in the reaction.

Nature of the Reactants

Depending upon what substances are reacting, the reaction rate varies. Acid/base reactions, the formation of salts, and ion exchange are fast reactions. When covalent bond formation takes place between the molecules and when large molecules are formed, the reactions tend to be very slow. Nature and strength of bonds in reactant molecules greatly influence the rate of its transformation into *products*.

Physical State

The physical state(solid, liquid, or gas) of a reactant is also an important factor of the rate of change. When reactants are in the same phase, as in aqueous solution, thermal motion brings them into contact. However, when they are in different phases, the reaction is limited to the interface between the reactants. Reaction can occur only at their area of contact; in the case of a liquid and a gas, at the surface of the liquid. Vigorous shaking and stirring may be needed to bring the reaction to completion. This means that the more finely divided a solid or liquid reactant the greater its surface area per unit volume and the more contact it makes with the other reactant, thus the faster the reaction. To make an analogy, for example, when one starts a fire, one uses wood chips and small branches — one does not start with large logs right away. In organic chemistry, on water reactions are the exception to the rule that homogeneous reactions take place faster than heterogeneous reactions.

Concentration

The reactions are due to collisions of reactant species. The frequency with which the molecules or ions collide depends upon their concentrations. The more crowded the molecules are, the more likely they are to collide and react with one another. Thus, an increase in the concentrations of the reactants will result in the corresponding increase in the reaction rate, while a decrease in the concentrations will have a reverse effect. For example, combustion that occurs in air (21% oxygen) will occur more rapidly in pure oxygen.

Temperature

Temperature usually has a major effect on the rate of a chemical reaction. Molecules at a higher temperature have more thermal energy. Although collision frequency is greater at higher temperatures, this alone contributes only a very small proportion to the increase in rate of reaction. Much more important is the fact that the proportion of reactant molecules with sufficient energy to react (energy greater than activation energy: $E > E_a$) is significantly higher and is explained in detail by the Maxwell–Boltzmann distribution of molecular energies.

The 'rule of thumb' that the rate of chemical reactions doubles for every 10 °C temperature rise is a common misconception. This may have been generalized from the special case of biological systems, where the a (temperature coefficient) is often between 1.5 and 2.5.

A reaction's kinetics can also be studied with a temperature jump approach. This involves using a sharp rise in temperature and observing the relaxation time of the return to equilibrium. A particularly useful form of temperature jump apparatus is a shock tube, which can rapidly jump a gas's temperature by more than 1000 degrees.

Catalysts

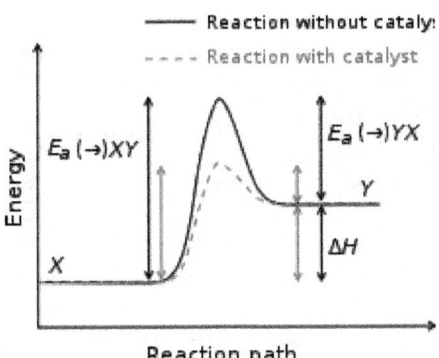

Reaction path

Generic potential energy diagram showing the effect of a catalyst in a hypothetical endothermic chemical reaction. The presence of the catalyst opens a different reaction pathway (shown in red) with a lower activation energy. The final result and the overall thermodynamics are the same.

A catalyst is a substance that accelerates the rate of a chemical reaction but remains chemically unchanged afterwards. The catalyst increases rate reaction by providing a different reaction mechanism to occur with a lower activation energy. In autocatalysis a reaction product is itself a catalyst for that reaction leading to positive to feedback. Proteins that act as catalysts in biochemical reactions are called enzymes. Michelis mentens kinetics describe the rate of enzyme mediated reactions. A catalyst does not affect the position of the equilibria, as the catalyst speeds up the backward and forward reactions equally.

In certain organic molecules, specific substituents can have an influence on reaction rate in neibouring group participation.

Agitating or mixing a solution will also accelerate the rate of a chemical reaction, as this gives the particles greater kinetic energy, increasing the number of collisions between reactants and, therefore, the possibility of successful collisions.

Pressure

Increasing the pressure in a gaseous reaction will increase the number of collisions between reactants, increasing the rate of reaction. This is because the activity of a gas is directly proportional to the partial pressure of the gas. This is similar to the effect of increasing the concentration of a solution.

In addition to this straightforward mass-action effect, the rate coefficients themselves can change due to pressure. The rate coefficients and products of many high-temperature gas-phase reactions change if an inert gas is added to the mixture; variations on this effect are called fall-off and chemical activation.

These phenomena are due to exothermic or endothermic reactions occurring faster than heat transfer, causing the reacting molecules to have non-thermal energy distributions (non-Boltzmann distribution). Increasing the pressure increases the heat transfer rate between the reacting molecules and the rest of the system, reducing this effect.

Condensed-phase rate coefficients can also be affected by (very high) pressure; this is a completely different effect than fall-off or chemical-activation. It is often studied using diamond anvils.

A reaction's kinetics can also be studied with a pressure jump approach. This involves making fast changes in pressure and observing the relaxation time of the return to equilibrium.

Equilibrium

While a chemical kinetics is concerned with the rate of a chemical reaction, thermodynamics determines the extent to which reactions occur. In a reversible reaction, chemical equilibrium is reached when the rates of the forward and reverse reactions are equal (the principle of detailed balance) and the concentrations of the reactants and products no longer change. This is demonstrated by, for example, the Haber–Bosch process for combining nitrogen and hydrogen to produce ammonia. Chemical clock reactions such as the Belousov–Zhabotinsky reaction demonstrate that component concentrations can oscillate for a long time before finally attaining the equilibrium.

Free Energy

In general terms, the free energy change (ΔG) of a reaction determines whether a chemical change will take place, but kinetics describes how fast the reaction is. A reaction can be very exothermic and have a very positive entropy change but will not happen in practice if the reaction is too slow. If a reactant can produce two different products, the thermodynamically most stable one will in general form, except in special circumstances when the reaction is said to be under kinetic reaction control. The Curtin–Hammett principle applies when determining the product ratio for two reactants interconverting rapidly, each going to a different product. It is possible to make predictions about reaction rate constants for a reaction from free-energy relationships.

The kinetic isotope effect is the difference in the rate of a chemical reaction when an atom in one of the reactants is replaced by one of its isotopes.

Chemical kinetics provides information on residence time and heat transfer in a chemical reactor in chemical engineering and the molar mass distribution in polymer chemistry.

Applications

The mathematical models that describe chemical reaction kinetics provide chemists and chemical engineers with tools to better understand and describe chemical processes such as food decomposition, microorganism growth, strato-spheric ozone decomposition, and the complex chemistry of biological systems. These models can also be used in the design or modification of chemical reactors to optimize product yield, more efficiently separate products, and eliminate environmentally harmful by-products. When performing catalytic cracking of heavy hydrocarbons into gasoline and light gas, for example, kinetic models can be used to find the temperature and pressure at which the highest yield of heavy hydrocarbons into gasoline will occur. Kinetics is also a basic aspect of chemistry.

Flow of Liquid Through a Hole

Stripping is a physical separation process where one or more components are removed from a liquid stream by a vapor stream. In industrial applications the liquid and vapor streams can have co-current or countercurrent flows. Stripping is usually carried out in either a packed or trayed column.

Theory

Stripping works on the basis of mass transfer. The idea is to make the conditions favorable for the component, A, in the liquid phase to transfer to the vapor phase. This involves a gas-liquid interface that A must cross. The total amount of A that has moved across this boundary can be defined as the flux of A, N_A.

Equipment

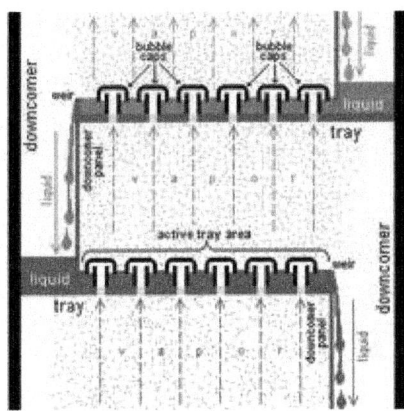

An example of a bubble cap tray that could be found inside of a stripping column.

Stripping is mainly conducted in trayed towers (plate columns) and packed columns, and less often in spray towers, bubble columns, and centrifugal contactors.

Trayed towers consist of a vertical column with liquid flowing in the top and out the bottom. The vapor phase enters in the bottom of the column and exits out of the top. Inside of the column are trays or plates. These trays force the liquid to flow back and forth horizontally while the vapor bubbles up through holes in the trays. The purpose of these trays is to increase the amount of contact area between the liquid and vapor phases.

Packed columns are similar to trayed columns in that the liquid and vapor flows enter and exit in the same manner. The difference is that in packed towers there are no trays. Instead, packing is used to increase the contact area between the liquid and vapor phases. There are many different types of packing used and each one has advantages and disadvantages.

Variables

The variables and design considerations for strippers are many. Among them are the entering conditions, the degree of recovery of the solute needed, the choice of the stripping agent and its flow, the operating conditions, the number of stages, the heat effects, and the type and size of the equipment.

The degree of recovery is often determined by environmental regulations, such as for volatile organic compounds like chloroform.

Frequently, steam, air, inert gases, and hydrocarbon gases are used as stripping agents. This is based on solubility, stability, degree of corrosiveness, and availability. As stripping agents are gases, operation at nearly the highest temperature and lowest pressure that will maintain the components and not vaporize the liquid feed stream is desired. This allows for the minimization of flow. As with all other variables, minimizing cost while achieving efficient separation is the ultimate goal.

The size of the equipment, and particularly the height and diameter, is important in determining the possibility of flow channeling that would reduce the contact area between the liquid and vapor streams. If flow channeling is suspected to be occurring, a redistribution plate is often necessary to, as the name indicates, redistribute the liquid flow evenly to reestablish a higher contact area.

As mentioned previously, strippers can be trayed or packed. Packed columns, and particularly when random packing is used, are usually favored for smaller columns with a diameter less than 2 feet and a packed height of not more than 20 feet. Packed columns can also be advantageous for corrosive fluids, high foaming fluids, when fluid velocity is high, and when particularly low pressure drop is desired. Trayed strippers are advantageous because of ease of design and scale up. Structured packing can be used similar to trays despite possibly being the same material as dumped (random) packing. Using structured packing is a common method to increase the capacity for separation or to replace damaged trays.

Trayed strippers can have sieve, valve, or bubble cap trays while packed strippers can have either structured packing or random packing. Trays and packing are used to increase the contact area over which mass transfer can occur as mass transfer theory dictates. Packing can have varying material, surface area, flow area, and associated pressure drop. Older generation packing include ceramic Raschig rings and Berl saddles. More common packing materials are metal and plastic Pall rings, metal Michael Bialecki rings, and ceramic Intalox saddles. Each packing material of this newer generation improves the surface area, the flow area, and/ or the associated pressure drop across the packing. Also important, is the ability of the packing material to not stack on top of itself. If such stacking occurs, it drastically reduces the surface area of the material. Lattice design work has been increasing of late that will further improve these characteristics.

During operation, monitoring the pressure drop across the column can help to determine the performance of the stripper. A changed pressure drop over a significant range of time can be an indication that the packing may need to be replaced or cleaned.

Typical Applications

Stripping is commonly used in industrial applications to remove harmful contaminants from waste streams. One example would be the removal of TBT and PAH contaminants from harbor soils.[2] The soils are dredged from the bottom of contaminated harbors, mixed with water to make a slurry and then stripped with steam. The cleaned soil and contaminant rich steam mixture are then separated. This process is able to decontaminate soils almost completely.

Steam is also frequently used as a stripping agent for water treatment. Volatile organic compounds are partially soluble in water and because of environmental considerations and regulations, must be removed from groundwater, surface water, and wastewater.[3] These compounds can be present because of industrial, agricultural, and commercial activity.

Flow Chemistry

In flow chemistry, a chemical reaction is run in a continuously flowing stream rather than in batch production. In other words, pumps move fluid into a tube, and where tubes join one another, the fluids contact one another. If these fluids are reactive, a reaction takes place. Flow chemistry is a well-established technique for use at a large scale when manufacturing large quantities of a given material. However, the term has only been coined recently for its application on a laboratory scale. Often, microreactors are used.

Batch *vs.* Flow

Comparing parameters in Batch *vs* Flow:

- Reaction stoichiometry. In batch production this is defined by the concentration of chemical reagents and their volumetric ratio. In Flow this is defined by the concentration of reagents and the ratio of their flow rate.
- Residence time. In batch production this is determined by how long a vessel is held at a given temperature. In flow the volumemetric residence time is used given by the ratio of volume of the reactor and the overall flow rate, as most often, plug flow reactors are used.

Running Flow Reactions

Choosing to run a chemical reaction using flow chemistry, either in a micro-reactor or other mixing device offers a variety of pros and cons.

Advantages

- Reaction temperature can raised above the solvent's boiling point as the volume of the laboratory devices is typically small. Typically, non-compressible fluids are used with no gas volume so that the expansion factor as a function of pressure is small.
- Mixing can be achieved within seconds at the smaller scales used in flow chemistry.
- Heat transfer is intensified. Mostly, because the area to volume ratio is large. Thereby, endothermal and exothermal reaction can be thermostated. The temperature gradient can be steep, allowing efficient control over reaction time.
- Safety is increased:
- Thermal mass of the system is dominated by the apparatus making thermal runaways unlikely.
- Smaller reaction volume is also considered a safety benefit.
- The reactor operates under steady-state conditions.
- Flow reactions can be automated with far less effort than batch reactions. This allows for unattended operation and experimental planning. By coupling the output of the reactor to a detector system, it is possible to go further and create an automated system which can sequentially investigate a range of possible reaction parameters (varying stoichiometry, residence time and temperature) and therefore explore reaction parameters with little or no intervention.

Typical drivers are higher yields/selectivities, less needed manpower or a higher safety level.

- Multi step reactions can be arranged in a continuous sequence. This can be especially beneficial if intermediate compounds are unstable, toxic, or sensitive to air, since they will exist only momentarily and in very small quantities.
- Position along the flowing stream and reaction time point are directly related to one another. This means that it is possible to arrange the system such that further reagents can be introduced into the flowing reaction stream at precisely the time point in the reaction that is desired.
- It is possible to arrange a flowing system such that purification is coupled with the reaction. There are three primary techniques that are used:
- Solid phase scavenging
- Chromatographic separation

- Liquid/Liquid Extraction
- Reactions which involve reagents containing dissolved gases are easily handled, whereas in batch a pressurized "bomb" reactor would be necessary.
- Multi phase liquid reactions (*e.g.* phase transfer catalysis) can be performed in a straightforward way, with high reproducibility over a range of scales and conditions.
- Scale up of a proven reaction can be achieved rapidly with little or no process development work, by either changing the reactor volume or by running several reactors in parallel, provided that flows are recalculated to achieve the same residence times.

Disadvantages

- Dedicated equipment is needed for precise continuous dosing (*e.g.* pumps), connections, *etc.*
- Start up and shut down procedures have to be established.
- Scale up of micro effects such as the high area to volume ratio is not possible and economy of scale may not apply. Typically, a scale up leads to a dedicated plant.
- Safety issues for the storage of reactive material still apply.

The drawbacks have been discussed in view of establishing small scale continuous production processes by Pashkova and Greiner.

Continuous Flow Reactors

Continuous reactors are typically tube like and manufactured from non-reactive materials such as stainless steel, glass and polymers. Mixing methods include diffusion alone (if the diameter of the reactor is small *e.g.* <1 mm, such as in microreactors) and static mixers. Continuous flow reactors allow good control over reaction conditincluding heat transfer, time and mixing.

The residence time of the reagents in the reactor (*i.e.* the amount of time that the reaction is heated or cooled) is calculated from the volume of the reactor and the flow rate through it:

Residence time = Reactor Volume / Flow Rate

Therefore, to achieve a longer residence time, reagents can be pumped more slowly and/or a larger volume reactor used. Production rates can vary from nano liters to liters per minute.

Some examples of flow reactors are spinning disk reactors (Colin Ramshaw); spinning tube reactors; multi-cell flow reactors; oscillatory flow reactors; microreactors; hex reactors; and 'aspirator reactors'. In an aspirator reactor a pump propels one reagent, which causes a reactant to be sucked in. This type of reactor was patented around 1941 by the Nobel company for the production of nitroglycerin.

Flow Reactor Scale

The smaller scale of micro flow reactors or microreactors can make them ideal for process development experiments. Although it is possible to operate flow processes at a ton scale, synthetic efficiency benefits from improved thermal and mass transfer as well as mass transport.

Fig. : A microreactor.

KEY APPLICATION AREAS

Use of Gases in Flow

Laboratory scale flow reactors are ideal systems for using gases, particularly those that are toxic or associated with other hazards. The gas reactions that have been most successfully adapted to flow are Hydrogenation and carbonylation although work has also been performed using other gases, *e.g.* ethylene and ozone.

Reasons for the suitability of flow systems for hazardous gas handling are:

- Systems allow the use of a fixed bed catalyst. Combined with low solution concentrations, this allows all compound to be adsorbed to catalyst in the presence of gas
- Comparatively small amounts of gas are continually exhausted by the system, eliminating the need for many of the special precautions normally required for handling toxic and/or flammable gases
- The addition of pressure means that a far greater proportion of the gas will be in solution during the reaction than is the case conventionally
- The greatly enhanced mixing of the solid, liquid and gaseous phases allows the researcher to exploit the kinetic benefits of elevated temperatures without being concerned about the gas being displaced from solution

Process Development

The process development change from a serial approach to a parallel approach. In batch the chemist works first followed by the chemical engineer. In flow chemistry this changes to a parallel approach, where chemist and chemical engineer work interactively. Typically there is a plant setup in the lab, which is the a tool for both. This setup can be either commercial or non commercial. The development scale can be small (ml/hour) for idea verification using a chip system and in the range of a couple of liters per hour for scalable systems like the flow miniplant technology. Chip systems are mainly used for liquid-liquid application while flow miniplant systems can deal with solids or viscous material.

Scale up of Microwave Reactions

Microwave reactors are frequently used for small scale batch chemistry. However due to the extremes of temperature and pressure reached in a microwave it is often difficult to transfer these reactions to conventional non-microwave apparatus for subsequent development, leading to difficulties with scaling studies. A flow reactor with suitable high temperature ability and pressure control can directly and accurately mimic the conditions created in a microwave reactor.This eases the synthesis of larger quantities by extending reaction time.

Manufacturing Scale Solutions

Flow systems can be scaled to the tons per hour scale. Plant redesign (batch to conti for an existing plant), Unit Operation (exchaning only one reaction step) and Modular Multi-purpose (Cutting a continuous plant into modular units) are typical realization solutions for flow processes.

Other uses of flow[shhttp://en.wikipedia.org/w/index.php?title=Flow_chemistry&action=edit&section=12].

It is possible to run experiments in flow using more sophisticated techniques, such as solid phase chemistries. Solid phase reagents, catalysts or scavengers can be used in solution and pumped through glass columns, for example, the synthesis of alkaloid natural product oxomaritidine using solid phase chemistries.[9]

There is an increasing interest in polymerization as a continuous flow process. For example **Reversible Addition -Fragmentation Chain Transfer** or **RAFT** polymerization.

Continuous flow techniques have also been used for controlled generation of nanoparticles. The very rapid mixing and excellent temperature control ofmicroreactors are able to give consistent and narrow particle size distribution of nanoparticles.

Segmented Flow Chemistry

As discussed above, running experiments in continuous flow systems is difficult, especially when one is developing new chemical reactions, which requires screening of multiple components, varying stoichiometry, temperature and residence time. In continuous flow, experiments are performed serially, which means one experimental condition can be tested. Experimental throughput is highly variable and as typically five times the residence time is needed for obtaining steady state. For temperature variation the thermal mass of the reactor as well as peripherals such as fluid baths need to be considered. More often than not, the analysis time needs to be considered.

Segmented flow is an approach that improves upon the speed in which screening, optimization and libraries can be conducted in flow chemistry. Segmented flow uses a "Plug Flow" approach where specific volumetric experimental mixtures are created and then injected into a high pressure flow reactor. Diffusion of the segment (reaction mixture) is minimized by using immiscible solvent on the leading and rear ends of the segment.

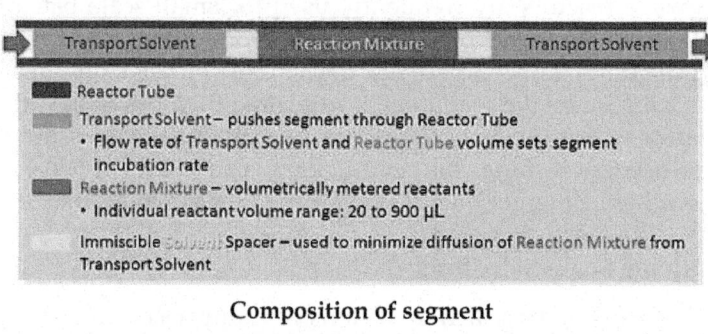

Composition of segment

Segment serial flow

One of the primary benefits of segmented flow chemistry is the ability to run experiments in a serial/parallel manner where experiments that share the same residence time and temperature can be repeatedly created and injected. In addition, the volume of each experiment is independent to that of the volume of the flow tube thereby saving a significant amount of reactant per experiment. When performing reaction screening and libraries, segment composition is typically varied by composition of matter. When performing reaction optimization, segments vary by stoichiometry.

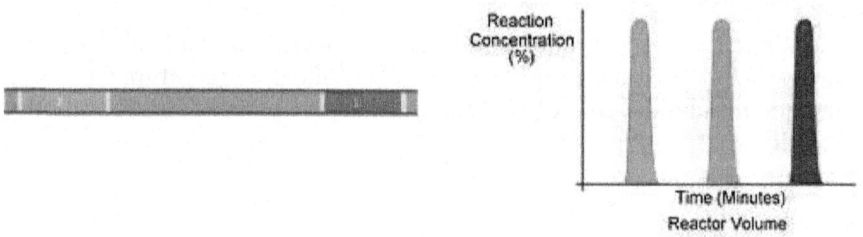

Segment serial/parallel Flow   Serial/Parallel Segments

Segmented flow is also used with online LCMS, both analytical and preparative where the segments are detected when exiting the reactor using UV and subsequently diluted for analytical LCMS or injected directly for preparative LCMS.

Flashing Liquids

Flash (or partial) evaporation is the partial vapor that occurs when a saturated liquidstream undergoes a reduction in pressure by passing through a throttling valve or other throttling device. This process is one of the simplest unit operations. If the throttling valve or device is located at the entry into a pressure vessel so that the flash evaporation occurs within the vessel, then the vessel is often referred to as a flash drum.[1][2]

If the saturated liquid is a single-component liquid (for example, liquid propane or liquid ammonia), a part of the liquid immediately "flashes" into vapor. Both the vapor and the residual liquid are cooled to the saturation temperature of the liquid at the reduced pressure. This is often referred to as "auto-refrigeration" and is the basis of most conventional vapor compression refrigeration systems.

If the saturated liquid is a multi-component liquid (for example, a mixture of propane,isobutane and normal butane), the flashed vapor is richer in the more volatile components than is the remaining liquid.

Uncontrolled flash evaporation can result in a boiling liquid expanding vapor explosion (BLEVE).

FLASH EVAPORATION OF A SINGLE-COMPONENT LIQUID

The flash evaporation of a single-component liquid is an isenthalpic process and is often referred to as an **adiabatic flash**. The following equation, derived from a simple heat balance around the throttling valve or device, is used to predict how much of a single-component liquid is vaporized.

$$X = \frac{H_u^L - H_d^L}{H_d^V - H_d^L}$$

where:

X = weight fraction vaporized

H_u^L = upstream liquid enthalpy at upstream temperature and pressure, J/kg

H_d^V = flashed vapor enthalpy at downstream pressure and corresponding saturation temperature, J/kg

H_d^L = residual liquid enthalpy at downstream pressure and corresponding saturation temperature, J/kg

If the enthalpy data required for the above equation is unavailable, then the following equation may be used.

$$X = \frac{c_p(T_u - T_d)}{H_v}$$

where:

X = weight fraction vaporized

C_p = liquid specific heat at upstream temperature and pressure, J/(kg °C)

T_u = upstream liquid temperature, °C

T_d = liquid saturation temperature corresponding to the downstream pressure, °C

H_v = liquid heat of vaporization at downstream pressure and corresponding saturation temperature, J/kg

Here, the words "upstream" and "downstream" refer to before and after the liquid passes through the throttling valve or device.

This type of flash evaporation is used in the desalination of brackish water or ocean water by "Multi-Stage Flash Distillation." The water is heated and then routed into a reduced-pressure flash evaporation "stage" where some of the water flashes into steam. This steam is subsequently condensed into salt-free water. The residual salty liquid from that first stage is introduced into a second flash evaporation stage at a pressure lower than the first stage pressure. More water is flashed into steam which is also subsequently condensed into more salt-free water. This sequential use of multiple flash evaporation stages is continued until the design objectives of the system are met. A large part of the world's installed desalination capacity uses multi-stage flash distillation. Typically such plants have 24 or more sequential stages of flash evaporation.

Equilibrium Flash of a Multi-component Liquid

The **equilibrium flash** of a multi-component liquid may be visualized as a simple distillation process using a single equilibrium stage. It is very different and more complex than the flash evaporation of single-component liquid. For a multi-component liquid, calculating the amounts of flashed vapor and residual liquid in equilibrium with each other at a given temperature and pressure requires a trial-and-error iterative solution. Such a calculation is commonly referred to as an equilibrium flash calculation. It involves solving the *Rachford-Rice equation*:

$$\sum_i \frac{z_i(1-K_i)}{1 + \beta(K_i - 1)} = 0$$

where:

- z_i is the mole fraction of component i in the feed liquid (assumed to be known);
- β is the fraction of feed that is vaporised;
- K_i is the equilibrium constant of component i.

The equilibrium constants K_i are in general functions of many parameters, though the most important is arguably temperature; they are defined as:

$$y_i = K_i x_i$$

where:

- x_i is the mole fraction of component i in liquid phase;
- y_i is the mole fraction of component i in gas phase.

Once the Rachford-Rice equation has been solved for β, the compositions x_i and y_i can be immediately calculated as:

$$x_i = \frac{z_i}{1 + \beta(K_i - 1)}$$

$$y_i = K_i x_i.$$

The Rachford-Rice equation can have multiple solutions for β, at most one of which guarantees that all x_i and y_i will be positive. In particular, if there is only one β for which:

$$\frac{1}{1 - K_{max}} = \beta_{min} < \beta < \beta_{max} = \frac{1}{1 - K_{min}}$$

then that β is the solution; if there are multiple such β's, it means that either $K_{max} < 1$ or $K_{min} > 1$, indicating respectively that no gas phase can be sustained (and therefore $\beta = 0$) or conversely that no liquid phase can exist (and therefore $\beta = 1$).

It is possible to use Newton's method for solving the above water equation, but there is a risk of converging to the wrong value of β; it is important to initialise the solver to a sensible initial value, such as $(\beta_{max} + \beta_{min})/2$ (which is however not sufficient: Newton's method makes no guarantees on stability), or, alternatively, use a bracketing solver such as the bisection method or the Brent method, which are guaranteed to converge but can be slower.

The equilibrium flash of multi-component liquids is very widely utilized in petroleum refineries, petrochemical and chemical plants and natural gas processing plants.

Contrast with Spray Drying

Spray drying is sometimes seen as a form of flash evaporation. However, although it is a form of liquid evaporation, it is quite different from flash evaporation.

In spray drying, a slurry of very small solids is rapidly dried by suspension in a hot gas. The slurry is first atomized into very small liquid droplets which are then sprayed into a stream of hot dry air. The liquid rapidly evaporates leaving behind dry powder or dry solid granules. The dry powder or solid granules are recovered from the exhaust air by using cyclones, bag filters or electrostatic precipitators.

Natural Flash Evaporation

Natural flash vaporization or flash deposition may occur during earthquakes resulting in depositing of minerals held in supersaturated solutions, sometimes even valuable ore in the case of auriferous, gold-bearing, waters. This results when blocks of rock are rapidly pulled and pushed away from each other by jog faults.

Flash evaporation is the partial or total vaporization that occurs when a saturated liquid stream undergoes a reduction in pressure by passing through a throttling valve or other throttling device. This process is one of the simplest unit operations. If the throttling valve or device is located at the entry into a pressure vessel so that the flash evaporation occurs within the vessel, then the vessel is often referred to as a flash drum.

If the saturated liquid is a single-component liquid (for example, liquid propane or liquid ammonia), a part of the liquid immediately "flashes" into vapor (*i.e.*, evaporates). Both the vapor and the residual liquid are cooled to the saturation temperature of the liquid at the reduced pressure. This is often referred to as "auto-refrigeration" and is the basis of most conventional vapor-compression refrigeration systems.

If the saturated liquid is a multi-component liquid (for example, a mixture of propane, isobutane and normal butane), a part of the liquid will also immediately flash into a vapor and the flashed vapor will be richer in the more volatile components than is the remaining liquid.

Flash Evaporation of a Single-component Liquid

The flash evaporation of a single-component liquid is an isenthalpic (*i.e.*, constant enthalpy) process and is often referred to as an adiabatic flash, a flash distillation or a throttling expansion. The following equation, derived from a simple heat balance around the throttling valve or device, is used to predict how much of a single-component liquid is vaporized when it "flashes":

$$X = 100 \, (H_u^L - H_d^L) \div (H_d^V - H_d^L)$$

where:

X = weight percent vaporized

H_u^L = upstream liquid enthalpy at upstream temperature and pressure, J/kg

H_d^V = flashed vapor enthalpy at downstream pressure and corresponding saturation temperature, J/kg

H_d^L = residual liquid enthalpy at downstream pressure and corresponding saturation temperature, J/kg

If the enthalpy data required for the above equation is unavailable, then the following equation may be used.

$$X = 100 \times c_p \, (T_u - T_d) \div H_v$$

where:

X = weight percent vaporized

c_p = liquid specific heat at upstream temperature and pressure, $J/(kg \cdot °C)$

T_u = upstream liquid temperature, °C

T_d = liquid saturation temperature corresponding to the downstream pressure, °C

H_v = liquid heat of vaporization at downstream pressure and corresponding saturation temperature, J/kg

(Note: The words "upstream" and "downstream" refer to before and after the liquid passes through the throttling valve or device.)

This type of flash evaporation is used in the desalination of brackish water or ocean water by Multi-Stage Flash Distillation. The water is heated and then routed into a reduced-pressure flash evaporation "stage" where some of the water flashes into steam. This steam is subsequently condensed into salt-free water. The residual salty liquid from that first stage is introduced into a second flash evaporation stage at a pressure lower than the first stage pressure. More water is flashed into steam which is also subsequently condensed into more salt-free water. This sequential use of multiple flash evaporation stages is continued until the design objectives of the system are met. A large part of the world's installed desalination capacity uses multi-stage flash distillation. Typically such plants have 24 or more sequential stages of flash evaporation.

Equilibrium Flash of a Multi-component Liquid

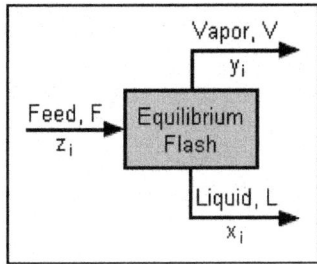

Fig. : Equilibrium flash of a multi-component feed

The equilibrium flash of a multi-component liquid is also an isenthalpic process and may be visualized as a simple distillation process using a single equilibrium stage. It is very different and more complex than the flash evaporation of single-component liquid. For a multi-component liquid, calculating the amounts of flashed vapor and residual liquid in equilibrium with each other at a given temperature and pressure requires a trial-and-error iterative solution. Such a calculation is commonly referred to as an equilibrium flash calculation. It involves solving the following Rachford-Rice equation:

$$\sum ni=1[zi(Ki-1)] \div [1+a(Ki-1)] = 0$$

where:

- z_i = mole fraction of component i in the feed liquid
- a = mole fraction of feed that is vaporized = V/F
- K_i = vapor-liquid equilibrium constant = y_i/x_i
- y_i = mole fraction of component i in the flashed vapor
- x_i = mole fraction of component i in the residual liquid

Newton's method (also known as the Newton-Raphson method) is an efficient iterative algorithm for solving the Rachford-Rice equation. Alternatively, the bisection method or the Brent method may be used. Once the equation has been solved for a, the compositions x_i and y_i can be immediately calculated as:

$$x_i = zi \div [1+a(K_i-1)]$$

$$y_i = K_i x_i$$

The equilibrium flash of multi-component liquids is very widely utilized in petroleum refineries, petrochemical and chemical plants and natural gas processing plants.

Liquid Pool Evaporation or Boiling

When a liquid is in contact with a surface maintained at a temperature above the saturation temperature of the liquid, boiling will eventually occur at that liquid-solid interface. Conventionally, based on the relative bulk motion of the body of a liquid to the heating surface, the boiling is divided into two categories; pool boiling and convective boiling.

Pool boiling is the process in which the heating surface is submerged in a large body of stagnant liquid. The relative motion of the vapor produced and the surrounding liquid near the heating surface is due primarily to the buoyancy effect of the vapor. Nevertheless, the body of the liquid as a whole is essentially at rest. Though the study on the boiling process can be traced back to as early as the eighteen century (the observation of the vapor film in the boiling of liquid over the heating surface by Leiden in 1756), the extensive study on the effect of the very large difference in the temperature of the heating surface and the liquid, ΔT, was first done by Nukiyama (1934). However, it was the experiment by Farber and Scorah (1948) that gave the complete picture of the heat transfer rate in the pool boiling process as a function of ΔT.

Farber and Scorah conducted their experiments by heating the water at various pressures with a heated cylindrical wire submerged horizontally under the water level. From the results, they divided the boiling curve into 6 regions based on the observable patterns of vapor production. Region I, ΔT is so small that the vapor is produced by the evaporation of the liquid into gas nuclei on the exposed surface of the liquid. Region II, ΔT is large enough that additional small bubbles are produced along the heating surface but later condense in the region above the superheated liquid. Region III, ΔT is enough to sustain "nucleate boiling", with the creation of the bubbles such that they depart and rise through the liquid regardless of the condensation rate. Region IV, an unstable film of vapor was formed over the heating surface, and oscillates due to the variable presence of the film. In this region, the heat transfer rate decreases due to the increased presence of the vapor film. Region V, the film becomes stable and the heat transfer rate reaches a minimum point. In Region VI, the ΔT is very large, and "film boiling" is stable such that the radiation through the film becomes significant and thus increases the heat transfer rate with the increasing ΔT.

This behavior as described above occurred when the temperature of the wire was the controlled parameter, If the power is the controlled variable then the increase in the power (or heat flux, q'') in Region III results in a jump in the wire surface temperature to a point in Region VI. This point of transition is known as the critical heat flux and occurs due to hydrodynamic fluid instabilities as discussed later. This results in the stable vapor film being formed, and the wire surface temperature increases as the heat transfer resistance increases for a fixed input power. If the power is now decreased, the vapor film remains stable in Region VI and the ΔT decreases to the minimum point for film boiling within Region V. At this point the vapor film becomes unstable and it collapses, with "nucleate boiling" becoming the mode of energy transfer. Thus, one passes quickly through Region IV and III to a lower wire surface temperature. This "hysteresis" behavior is always seen when the power (or heat flux) is the controlled parameter.

On the effect of the pressure, Farber and Scorah suggested that increasing the pressure, for the same temperature difference, would result in the decreasing of the size of the bubbles. At the same time, the film becomes thinner and less circulation would be observed. This effect is counter balanced by the increased density of the vapor and the attendant increase in its enthalpy. Thus, the increase in pressure initially increased the heat transfer rate in pool boiling. The objective of this section is to present an overall picture of the pool boiling process with an emphasis on the practical models used (1) to identify the transition between natural convection and nucleate boiling, as well as nucleate and film boiling, and (2) to estimate the heat flux during nucleate and film boiling given the difference between the heater surface temperature and the bulk liquid. First, we consider the process of bubble nucleation and then begin to "construct" the conceptual picture of the pool boiling curve with suggested quantitative models.

Accidental Release of a Pressurized Gas

When gas stored under pressure in a closed vessel is discharged to the atmosphere through a hole or other opening, the gas velocity through that opening may be choked or it may be non-choked. Choked flow (also referred to as "critical flow") is a limiting or maximum condition at which the gas velocity has attained the speed of sound in the gas.

Choked flow occurs when the ratio of the absolute upstream pressure to the absolute downstream pressure is equal to or greater than:

(1) $[(k+1)/2]^{k/(k-1)}$

where k is the specific heat ratio of the discharged gas (sometimes called the isentropic expansion factor and sometimes denoted as γ).

For many gases, k ranges from about 1.09 to about 1.41, and therefore the expression in (1) ranges from 1.7 to about 1.9, which means that choked velocity usually occurs when the absolute upstream vessel pressure is at least 1.7 to 1.9 times as high as the absolute downstream pressure. In the case of a leak to the ambient atmosphere, the downstream pressure is the atmospheric pressure.

When the gas velocity is choked, the equation for the mass flow rate in SI units is:

(2) $\dot{m} = C A \sqrt{k \rho_u P_u \left(\dfrac{2}{k+1}\right)^{(k+1)/(k-1)}}$

where the terms are defined as stated below. If the upstream gas density, ρ_u is not known directly, then it is useful to eliminate it by using the ideal gas law corrected for the real gas compressibility:

(3) $\dot{m} = C A P_u \sqrt{\left(\dfrac{k M}{Z R T_u}\right)\left(\dfrac{2}{k+1}\right)^{(k+1)/(k-1)}}$

For the above equations, it is important to note that although the gas velocity reaches a maximum and becomes choked, the mass flow rate is not choked. The mass flow rate can still be increased if the upstream pressure is increased or the temperature is decreased.

Whenever the ratio of the absolute upstream pressure to the absolute downstream pressure is less than in expression (1) above, then the gas velocity is non-choked and the equation for mass flow rate is:

(4) $\dot{m} = C A \sqrt{2 \rho_u \, P_u \left(\dfrac{k}{k-1}\right)\left[\left(\dfrac{P_d}{P_u}\right)^{2/k} - \left(\dfrac{P_d}{P_u}\right)^{(k+1)/k}\right]}$

or this equivalent form

(5) $\dot{m} = C A P_u \sqrt{\left(\dfrac{2 M}{Z R T_u}\right)\left(\dfrac{k}{k-1}\right)\left[\left(\dfrac{P_d}{P_u}\right)^{2/k} - \left(\dfrac{P_d}{P_u}\right)^{(k+1)/k}\right]}$

- where:
- $\dot{m}$ = mass flow rate, kg/s
- C = discharge coefficient, dimensionless (usually about 0.72)
- A = discharge hole area, m²
- k = cp ÷ cv = specific heat ratio of the gas
- cp = specific heat capacity of the gas at constant pressure

- cv = specific heat capacity of the gas at constant volume
- ρu = real gas upstream density, kg/m³ = (M Pu) ÷ (Z R Tu)
- Pu = absolute upstream pressure, Pa
- Pd = absolute downstream pressure, Pa
- M = the gas molecular mass, kg/kmol (also known as the molecular weight)
- R = the universal gas law constant = 8314.472 Pa·m³ ÷ (kmol·K)
- Tu = absolute upstream gas temperature, K
- Z = the gas compressibility factor at Pu and Tu, dimensionless

The above equations calculate the initial instantaneous mass flow rate for the pressure and temperature existing in the source vessel when a release first occurs. The initial instantaneous flow rate from a leak in a pressurized gas system or vessel is much higher than the average flow rate during the overall release period because the pressure and flow rate decrease with time as the system or vessel empties. Calculating the flow rate versus time since the initiation of the leak is much more complicated, but more accurate. A comparison between two methods for performing such calculations is available online.

The technical literature can be confusing because many authors do not explain whether they are using the universal gas law constant *R* which applies to any ideal gas or whether they are using R_s which only applies to a specific individual gas. The relationship between the two is $R_s = R/M$.

Notes:

- The above equations are for a real gas.
- For an ideal gas, Z = 1 and ρ is the ideal gas density.
- kmol = 1000 mol

Evaporation of a Non-boiling Liquid Pool

Three different methods of calculating the rate of evaporation from a non-boiling liquid pool are presented in this section. The results obtained by the three methods are somewhat different.

The U.S. EPA Method

The following equations are for predicting the rate at which liquid evaporates from the surface of a pool of liquid which is at or near the ambient temperature. The equations were developed by the U.S. Environmental Protection Agency using units which were a mixture of metric usage and United States usage. The non-metric units have been converted to metric units for this presentation.

- E = (0.1268 ÷ T) u0.78 M0.667 A P
- where:
- E = evaporation rate, kg/min

- u = windspeed just above the pool liquid surface, m/s
- M = pool liquid molecular mass, dimensionless
- A = surface area of the pool liquid, m^2
- P = vapor pressure of the pool liquid at the pool temperature, kPa
- T = pool liquid absolute temperature, K

The U.S. EPA also defined the pool depth as 0.01 m (*i.e.*, 1 cm) so that the surface area of the pool liquid could be calculated as:

- A = (pool volume, in m3) ÷ (0.01)

The U.S. Air Force Method

The following equations are for predicting the rate at which liquid evaporates from the surface of a pool of liquid which is at or near the ambient temperature. The equations were derived from field tests performed by the U.S. Air Force with pools of liquid hydrazine.

- E = (4.161 × 10– 5) u0.75 TF M (PS ÷ PH)
- where:
- E = evaporation flux, (kg/min) / m2 of pool surface
- u = windspeed just above the liquid surface, m/s
- TA = absolute ambient temperature, K
- TF = pool liquid temperature correction factor, dimensionless (see equations [a] and [b] below)
- TP = pool liquid temperature, °C
- M = pool liquid molecular weight, g/mol
- PS = pool liquid vapor pressure at ambient temperature, mmHg
- PH = hydrazine vapor pressure at ambient temperature, mmHg (see equation [c] below)
- [a] If TP = 0 °C or less, then TF = 1.0
- [b] If TP > 0 °C, then TF = 1.0 + 0.0043 TP2
- [c] PH = 760 exp[65.3319 – (7245.2 ÷ TA) – (8.22 ln TA) + (6.1557 × 10– 3) TA]

Note: The function ln x is the natural logarithm (base e) of x, and the function exp x is e (approximately 2.7183) raised to the power of x.

Stiver and Mackay's Method

The following equations are for predicting the rate at which liquid evaporates from the surface of a pool of liquid which is at or near the ambient temperature. The equations were developed by Warren Stiver and Dennis Mackay of the Chemical Engineering Department at the University of Toronto.

- $E = k P M \div (R\, TA)$
- where:
- E = evaporation flux, (kg/s)/m2 of pool surface
- k = mass transfer coefficient, m/s (which is taken to be 0.002 u)
- TA = absolute ambient temperature, K
- M = pool liquid molecular weight, g/mol
- P = pool liquid vapor pressure at ambient temperature, Pa
- R = the universal gas law constant of 8314.472 Pa ·m3 ÷ (kmol ·K)
- u = windspeed just above the liquid surface, m/s

Evaporation of Boiling, Cold Liquid Pool

The following equation is for predicting the rate at which liquid evaporates from the surface of a pool of cold liquid (*i.e.*, at a liquid temperature of about 0 °C or less).

- $E = (0.0001) (7.7026 - 0.0288\, B) (M)\, e\text{-}\, 0.0077B - 0.1376$

where:

- E = evaporation flux, (kg/min)/m2 of pool surface
- B = pool liquid atmospheric boiling point, °C
- M = pool liquid molecular weight, g/mol
- e = 2.7183, the base of the natural logarithm

Adiabatic Flash of Liquified Gas Release

Liquified gases such as ammonia or chlorine are often stored in cylinders or vessels at ambient temperatures and pressures well above atmospheric pressure. When such a liquified gas is released into the ambient atmosphere, the resultant reduction of pressure causes some of the liquified gas to vaporize immediately. This is commonly referred to as "adiabatic flashing" and the following equation, derived from a simple heat balance, is used to predict how much of the liquified gas is vaporized:

- $X = 100\, (HuL - HdL) \div (HdV - HdL)$
- where:
- X = weight percent vaporized
- HuL = upstream liquid enthalpy at upstream temperature and pressure, J/kg
- HdV = flashed vapor enthalpy at downstream pressure and corresponding saturation temperature, J/kg
- HdL = residual liquid enthalpy at downstream pressure and corresponding saturation temperature, J/kg

If the enthalpy data required for the above equation is unavailable, then the following equation may be used:

- $X = 100 \times cp \, (Tu - Td) \div Hv$
- where:
- X = weight percent vaporized
- cp = liquid specific heat at upstream temperature and pressure, J/(kg·°C)
- Tu = upstream liquid temperature, °C
- Td = liquid saturation temperature corresponding to the downstream pressure, °C
- Hv = liquid heat of vaporization at downstream pressure and corresponding saturation temperature, J/kg

Note: The words "upstream" and "downstream" refer to before and after the liquid passes through the release opening.

Chapter 5

A MODIFIED SURFACE ON TITANIUM DEPOSITED BY A BLASTING PROCESS

Caroline O'Sullivan[1,2,*], Peter O'Hare[3], Greg Byrne[4], Liam O'Neill[5],
Katie B. Ryan[1] and Abina M. Crean[1]

[1] School of Pharmacy, Cavanagh Building, University College Cork, Cork, Ireland;
E-Mails: katie.ryan@ucc.ie (K.B.R.); a.crean@ucc.ie (A.M.C.)

[2] Department of Chemical and Process Engineering, Cork Institute of Technology, Bishopstown, Cork, Ireland

[3] The Nanotechnology and Integrated BioEngineering Centre, University of Ulster at Jordanstown, Newtownabbey, Co Antrim, BT37 OQB, Northern Ireland; E-Mail: p.ohare@ulster.ac.uk

[4] School of Electrical, Electronic & Mechanical Engineering, University College Dublin, Belfield, Dublin 4, Ireland; E-Mail: gregory.byrne@ucd.ie

[5] Research & Development, EnBIO, Carrigtohill, Cork, Ireland; E-Mail: liam.oneill@enbio-materials.com

[*] Author to whom correspondence should be addressed; E-Mail: caroline.osullivan@cit.ie;
Tel.: +353-(0)21-490-1667; Fax: +353-(0)21-490-1656.

ABSTRACT

Hydroxyapatite (HA) coating of hard tissue implants is widely employed for its biocompatible and osteoconductive properties as well as its improved mechanical properties. Plasma technology is the principal deposition process for coating HA on bioactive metals for this application. However, thermal decomposition of HA can occur during the plasma deposition process, resulting in coating variability in terms of purity, uniformity and crystallinity, which can lead to implant failure caused by aseptic loosening. In this study, CoBlast™, a novel blasting process has been used to successfully modify a titanium (V) substrate with a HA treatment using a dopant/abrasive regime. The impact of a series of apatitic abrasives under the trade name MCD, was investigated to determine the effect of abrasive particle size on the surface properties of both microblast (abrasive only) and CoBlast (HA/abrasive) treatments. The resultant HA treated substrates were compared to substrates treated with abrasive only (microblasted) and an untreated Ti. The HA powder, apatitic abrasives and the treated substrates were characterized for chemical composition, coating coverage, crystallinity and topography including surface roughness. The results show that the surface roughness of the HA blasted modification was affected by the particle size of the apatitic abrasives used. The CoBlast process did not alter the chemistry of the crystalline HA during deposition. Cell proliferation on the HA surface was also assessed, which demonstrated enhanced osteo-viability compared to the microblast and blank Ti. This study demonstrates the ability of the CoBlast process to deposit HA coatings with a range of surface properties onto Ti substrates. The ability of the CoBlast technology to offer diversity in modifying surface topography offers exciting new prospects in tailoring the properties of medical devices for applications ranging from dental to orthopedic settings.

Keywords

Hydroxyapatite; grit blasting; CoBlast; hard tissue implants.

1. INTRODUCTION

Hydroxyapatite (HA), $Ca_{10}(PO_4)_6(OH)_2$, a proven bioceramic for coating medical device implants is widely known, not only for its biocompatible and osteoconductive properties, but also for its increased mechanical properties when applied to bio-inert metals for orthopedic use [1-4]. Implant surface modifications are often required in order to prescribe a particular surface roughness and increase surface area for osteoblast attachment, as well as to enhance the bioactive and osteoconductive properties of the underlying substrate. Such surface treatment methods include sand- or grit-blasting using abrasives, chemical treatments and deposition of calcium phosphate (CaP) coatings [2-8].

Abrasive blasting involves impacting the implant metal surface with abrasive particles under pressure to roughen the surface. Roughening orthopedic and dental implants utilizing alumina (Al_2O_3) abrasives is a common practice to enhance implant osteointegration in vivo [5,6]. However, the use of apatite abrasives are often preferred as it enhances bone formation [7,18]. It has been shown that this technique can be effective in depositing a thin layer of CaP on the surface being roughened [18-20]. A number of other HA coating deposition techniques have been employed to confer a bioactive layer onto metallic and other inert substrates including plasma spraying, which is one of the most common types of coating process for the generation of CaP thin films [3,4,9-14] and alternative deposition processes including pulsed laser deposition (PLD) [15], radio frequency (RF) magnetron sputtering [16], sol-gel immersion techniques, and electrophoretic deposition [17].

More recently, a novel approach CoBlast has been shown as an alternative process to deposit HA and substituted apatites onto titanium (Ti) substrates [21-23].

The CoBlast technique is based on the convergent flow of an abrasive and a dopant stream onto the implant surface which can effectively impregnate the metal with the dopant material. The CoBlast approach manipulates the ability of abrasive blasting to achieve surface roughening and bioactive layer deposition. The impregnation of the dopant material onto the surface results from a combination of the mechanical interlocking and tribo-chemical bond formation between the bioceramic material and the underlying metal substrate [21]. HA coatings prepared using the CoBlast technique demonstrated enhanced osteoblast attachment in vitro and early stage lamellar bone growth *in vivo* compared to microblasted and untreated Ti surfaces [21]. Additionally, a series of substituted apatites (AgA, SrA, ZnA) were effectively deposited using the CoBlast technique and these modifications offered the dual benefits of osteoconductive properties essential for bone integration with the added potential of microbial colonization inhibition without cytotoxic effects [23]. The established research showed that <10 μm thick coatings were applied with this technique employing alumina as the abrasive and that there was no evidence of alumina being incorporated into the modified surface [21].

The objective of this study is to demonstrate the use of apatitic abrasives in the treatment of Ti substrates using both the CoBlast technique (dopant/abrasive regime) and a control microblast surface (abrasive only). The chemical, topological and osteo-viability advantages of treated Ti substrates was characterised. The effect of abrasive particle size on the properties and performance of the CoBlast and microblast modified surfaces was also investigated. A series of apatitic abrasives (sintered CaP under the trade name MCD) with differing mean particle size values were employed for both techniques.

2. RESULTS AND DISCUSSION

2.1. Chemical Characterization of HA and MCD Abrasive Powders

The particle size of the HA and MCD abrasives were measured using a laser light technique (Mastersizer S), Table 1. The average particle size (d (0.5)) increased in the following order: HA < MCD-106 < MCD-180 < MCD-425. The various powders were analyzed for their chemical composition using energy dispersive X-ray (EDX) analysis, Table 1.

Table 1. Mean particle size analysis and energy dispersive X-ray (EDX) analysis of the calcium phosphate powders (n = 3).

Powder	Mean particle Size (μm)	O % atm	Ca % atm	P % atm	Ca/P
HA	40 (±4)	71	18	11	1.66
MCD-106	44 (±2)	72	18	10	1.76
MCD-180	124 (±6)	73	17	10	1.73
MCD-425	355 (±6)	77	13	10	1.29

The calcium phosphate powders (HA and MCD abrasives) were found to be composed of O, P and Ca. The Ca/P ratio for stoichiometric HA was found to be similar to the previously reported value of 1.67 [25]. The increase in Ca/P ratio for MCD-106 and MCD-180, as seen in Table 1, may be explained by the presence of impurities such as tricalcium phosphate (TCP) phase as determined by powder X-ray diffraction (PXRD) analysis (Figure 1). However, the Ca deficient nature for the more amorphous MCD-425 results in a reduced Ca/P ratio (1.29).

Relative crystallinity of each powder was investigated using PXRD, (Figure 1). HA was found to be highly crystalline with well defined narrow peaks. The main characteristic peaks associated with HA can be assigned to the 002, 102, 210, 211, 112, 300 and 202 reflections corresponding to 25.9°, 28.1°, 28.9°, 31.9°, 32.2°, 33.1° and 34.1°, as previously reported [25]. The resulting PXRD patterns for the MCD apatite series indicate a lower crystallinity relative to the HA powder. The small peak present at 31° and 34.4° was attributed to the TCP phase [26]. Also in the MCD-425 pattern, peaks are poorly resolved with low intensity relative to the other apatites, demonstrating the more amorphous nature of this material [25,26].

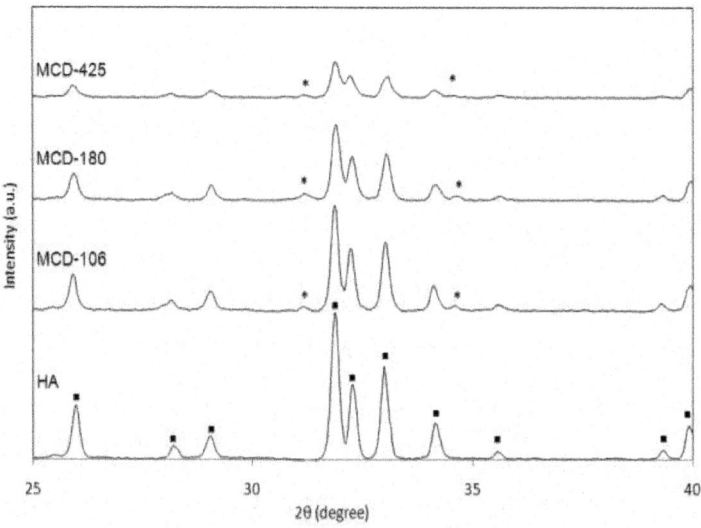

Figure 1. Powder X-ray diffraction (PXRD) spectra of the powders (■ denotes HA peaks and * represents tricalcium phosphate (TCP)).

The Fourier transform infrared spectrometer (FTIR) spectra of the powders in the range 1600–450 cm⁻¹ are presented in Figure 2. The most intense peaks observed for the crystalline HA powder are those attributable to vibrations of the PO_4^{3-} groups; the v_1 and v_3 phosphate bands in the region of 900–1200 cm⁻¹ and v_4 absorption bands in the region of 500–700 cm⁻¹, which are used to characterize apatite structure. The peak at 962 cm⁻¹ is assigned to the v_1 symmetric P-O stretching vibration of the PO_4^{3-} and the v_3 asymmetric P-O stretching mode are indexed at 1090 and 1045 cm⁻¹ [27]. The bands at 601 and 571 cm⁻¹ are assigned to v_4 vibration mode of the phosphate group, which occupies two crystal lattice sites (O-P-O bending mode) according to previous studies [27]. The HA adsorption bands of the v_1 and v_4 of the PO_4^{3-} groups determined here are those of stoichiometric HA [16]. The bands at 631 and 474 cm^{-1} correspond to the vibrations of OH$^-$ groups in the structure [27]. The MCD series showed similar finger-print bonds for calcium phosphate bonds but with broader definition, which is representative of the increased amorphous content of these CaP materials [28].

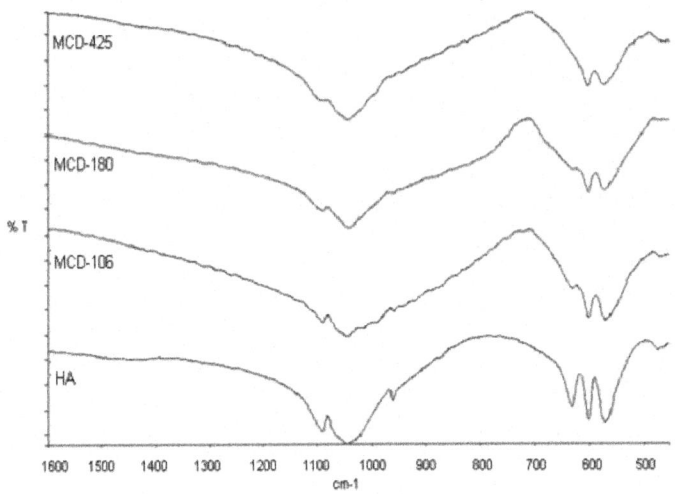

Figure 2. Fourier transform infrared spectrometer (FTIR) spectra of the various apatite powders.

2.2. Characterization of the Modified Titanium Substrates

Titanium (V) was used as the base substrate and the untreated Ti surface was determined to contain 23% O and 77% Ti using EDX analysis. The chemical composition of the microblast surfaces (abrasive blast only, no dopant) are presented in Table 2 and were analyzed for O, Ca, P and Ti only.

Table 2. EDX analysis, coating thickness (PosiTector thickness gauge) and mass of the modified surfaces.

Modification		O % atm	P % atm	Ca % atm	Ti % atm	Ca/P	Coating Thickness (µm) (2STD)	Coating Mass (mg/cm²) (2STD)
Blank	Ti	23	-	-	77	-	0	-
Microblast	MCD-106	59	8	12	21	1.56	3 ± 1	-
	MCD-180	56	6	10	27	1.53	3 ± 1	-
	MCD-425	55	7	11	27	1.57	3 ± 2	-
CoBlast	HA/MCD-106	63	13	21	2	1.59	6 ± 3	0.48 ± 0.4
	HA/MCD-180	67	12	18	3	1.53	6 ± 1	0.44 ± 0.4
	HA/MCD-425	65	12	20	5	1.61	7 ± 3	0.44 ± 0.3

The % atm Ti determined reflects the exposure of the underlying substrate and can be used to represent the degree of coverage resulting from apatite materials. A high Ti level represents a thin or patchy coating and conversely, a low Ti concentration signifies a thick coating. The EDX results reveal that after a wash treatment, the MCD microblasted surfaces show a reduction in Ti concentrations to 21–27% atm, compared to 77% atm for the Ti substrate. Also the presence of Ca and P which are the main constituents of the MCD abrasive is noted on the surface. This illustrates that a thin coating of Ca/P material has been blasted onto the surface and successfully deposited as a stable layer onto the Ti surface. The Ca/P values obtained for the microblasted samples treated with the MCD series of the abrasives ranged between 1.53 and 1.57 which is consistent with similar grit blasted studies [18].

The chemical composition of the CoBlast surfaces (blasting with both abrasive and dopant) on Ti are also presented in Table 2. For CoBlast coatings, the levels of O, P, Ca and Ti obtained were determined to be in the range of 63–67%, 12–13%, 18–21% and 2–5% atm, respectively. The reduced level of Ti detected in these samples, compared to the Ti substrate and microblasted surfaces, is indicative of a high degree of coating coverage across all CoBlast samples. The Ca/P values were found to display a ratio of between 1.53 and 1.61, which are relatively close to the value for stoichiometric HA [25]. The % atm Ti, determined using EDX analysis, was observed to increase as the MCD series particle size order increased, indicating a decrease in the thickness of the deposited layer, as outlined in Table 2. This suggests that the smaller the particle size of the MCD abrasive the more HA was deposited, although the coating thickness determined using the PosiTector thickness gauge, and the coating mass values were found to be similar. The coating thickness of all the CoBlast samples was <10 microns which is in agreement with a previous study which used Al_2O_3 as the abrasive [22].

The scanning electron microscopy (SEM) image of the untreated Ti substrate can be seen in Figure 3a which has similar topography to that observed in a previous study [6]. This image reveals a very smooth surface and the morphology of a machined metal. The SEM images of the microblast MCD-106 surfaces, as well as the corresponding CoBlast HA/MCD-106 surfaces, are presented in Figure 3b and c respectively. (More images can be seen as supporting information)

As expected, the microblast process was observed to roughen the untreated Ti surface. Examination of the CoBlast surfaces suggests that the co-introduction of the HA with the abrasive appears to have in-filled some of the surface features that are evident on the microblast sample (Figure 3b). The CoBlast process results in a roughened, highly regular and uniform surface which is consistent with other calcium phosphate coatings produced using simple grit blasting technologies [6,19,20]. It was noted that as the particle size (d90) of the MCD abrasive increased from 106 to 425 microns, the texture (presence of surface features) and the apparent roughness of the resultant surfaces was also observed to increase for both the microblast and CoBlast treatments and this was confirmed by surface roughness measurements.

(a)

(b)

(c)

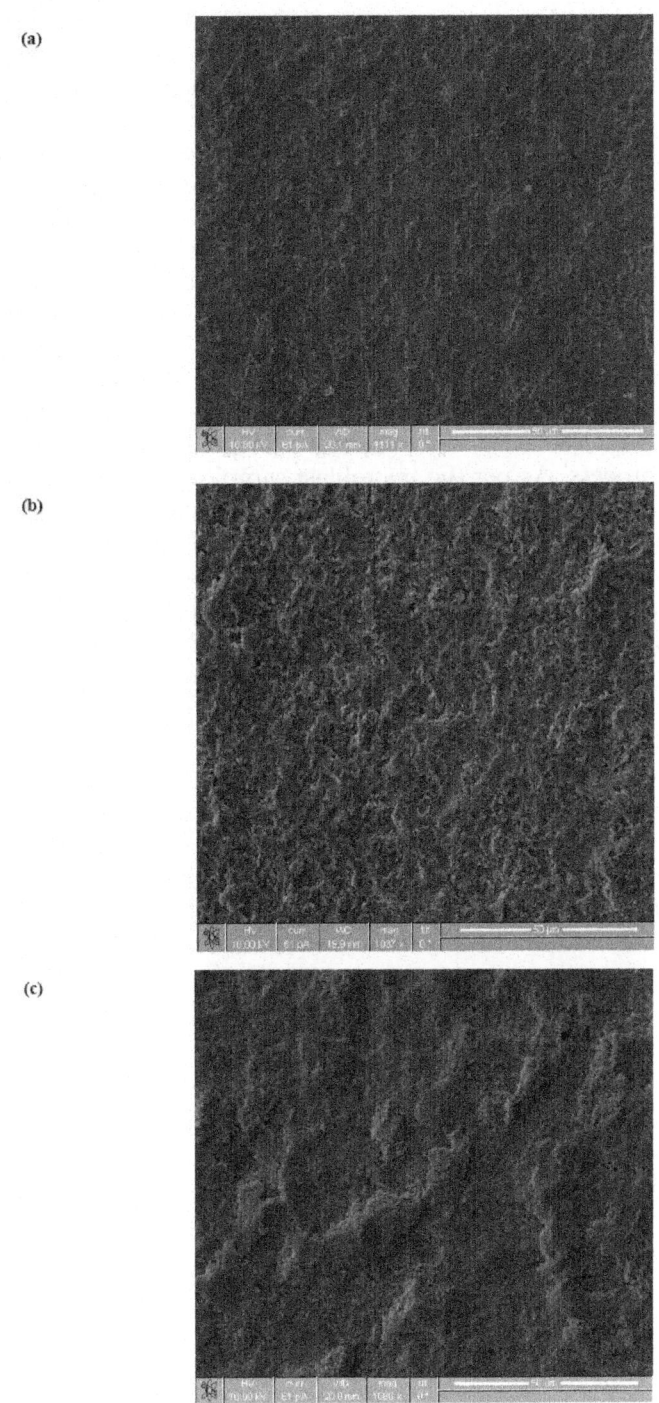

Figure 3. Scanning electron microscopy (SEM) images (×1000 magnification) of (a) titanium; (b) microblast MCD 106; (c) CoBlast HA/MCD-106.

The surface oughness was measured using a stylus method and the results obtained are given in Figure 4. The arithmetical mean roughness (R_a) was used as a measure of the surface roughness, which tended to increase as the particle size of the abrasive increased. Significant differences were observed between the surface treated and the untreated Ti. The average surface roughness of the blank titanium (0.4 μm) increased to 0.5, 0.8, 1.4 μm when MCD-106, MCD-180 and MCD-425, respectively were employed for microblast treatments. It has been previously reported that an increase in surface roughness was observed with the introduction of HA with the Al_2O_3 abrasive using the CoBlast technique and the same trend was observed here on the introduction of HA with the MCD abrasives [21]. Statistically significant differences were noted between the roughness of the microblast and CoBlast samples prepared using MCD-106 and MCD-180 abrasives, though not when MCD-425 was used. A large standard error was observed for the MCD-425 microblasted surface, which may possibly be a feature of the crude microblast process. As per the microblast samples, the level of roughness and irregularity of the CoBlast surface was visibly altered by changes in the abrasive particle size, with a significant increase in surface roughness produced by larger abrasive particle sizes ($p < 0.05$).

Implants with rougher surfaces result in a higher removal torque force and demonstrate excellent osteointegration when compared to those with smoother surfaces [5]. As seen in this study, microblasting offers increased roughening of machined Ti substrates as expected and the use of MCD abrasives results in the deposition of a thin coating layer of calcium phosphate. Furthermore, the roughness of the microblast process can be tuned between 0.5–1.4 μm depending on the particle size of the apatite abrasive employed. This is consistent with previously studies [5,6]. Unfortunately, the coatings deposited using this microblast process have demonstrated poor adhesion to the metal and have not been widely employed as final surface treatments for this reason [21].

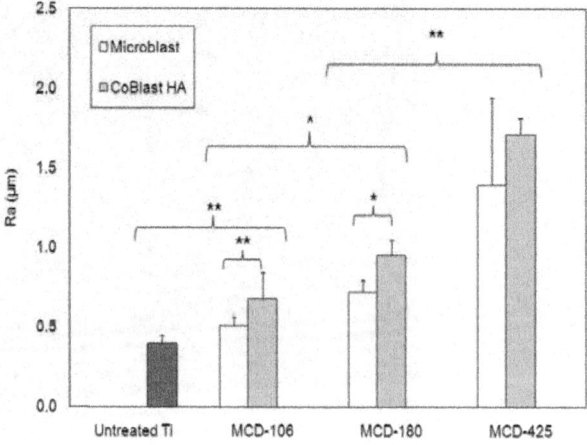

Figure 4. Surface roughness (R_a) of the various modifications (* denotes $p < 0.05$ and ** ascribes $p < 0.01$ determined using student's t-test).

For the CoBlast samples, the tribo-chemical bonding which results from surface roughening, activation and subsequent bonding of the powder to the substrate has been shown to improve the HA bonding to the titanium [21,22]. The deposition of HA via the CoBlast process combines the benefits of increased roughness and enhanced bioceramic deposition with the added bioactive property of improved osteointegration compared to the microblasted surfaces. The surface treatment effect was also dependant on the particle size of the abrasive, with the larger particle size producing greater surface erosion and a rougher topography resulting in reduced coating thickness.

In literature, HA coatings, deposited using the standard high temperature plasma spray deposition technique, have been reported to contain a variety of crystalline phases and the presence of altered chemical functionality [9]. This can lead to the formation of intermediates such as calcium oxide (CaO), octa-hydroxya-patite (OHA), α-tricalcium phosphate (α-TCP), β-tricalcium phosphate (β-TCP) and tetra-calcium phosphate (TTCP) [10,11]. The presence of these impurities in a HA modified surface can decrease the crystallinity and subsequently make it more prone to dissolution. The increased solubility of these phases can eventually lead to the poor apposition of bone and compromize the mechanical stability of the implant [12, 29]. This can occur when the coating itself de-laminates over time due to poor bonding strength of the HA onto the underlying surface or as the coating itself resorbs into the surrounding environment [30].

Figure 5 shows the XRD pattern of the CoBlast HA coated substrates. Due to the thin nature of the deposited material and the interference of the background Ti metal, detailed analysis of the XRD profiles was not possible. However, for the CoBlast substrates, the peaks detected clearly correspond to that of HA powder employed. The additional 2θ peaks observed at 35.3° and 38.5° were assigned

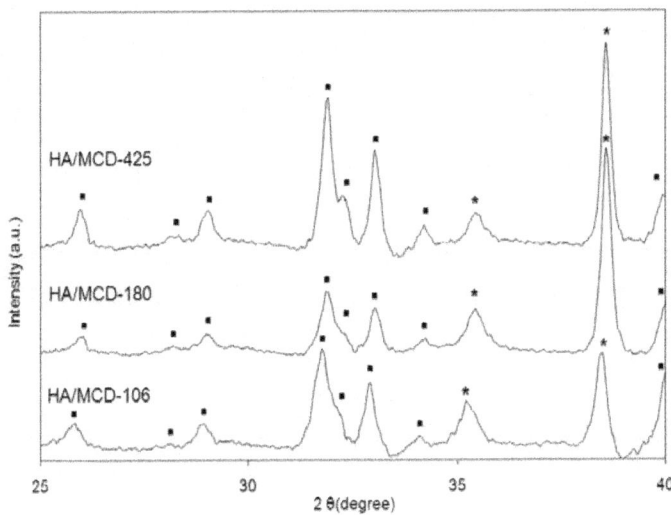

Figure 5. XRD of the CoBlast HA modifications (■ denotes HA peaks;*denotes Ti peaks).

to the Ti substrate. No evidence of the TCP phase was detected (31° and 34.4°), which suggests that no compositional or crystallographic changes occurred to the HA powder during the blasting process, with negligible uptake of the abrasive, which is in keeping with previous studies [21,22].

Figure 6 shows the FTIR spectrum in the range 1600–450 cm^{-1} for the HA coated substrates. The FTIR spectra for all CoBlast HA coatings irrespective of the abrasive used are very similar and display the characteristic features of the HA powder used, as discussed earlier. There was no evidence of hydration (broad banding at 3450 cm^{-1}), further demonstrating a pure HA coating has been deposited. Also, the position of the characteristic peak at 962 cm^{-1} represents a highly ordered, non-carbonated apatite and indicates a highly crystalline nature [16]. The banding assignments are in agreement with those of the HA powder as seen in Figure 2 suggesting that the chemistry is retained during the deposition process. This also supports the XRD analysis discussed above, which suggests that there has been minimal uptake of the abrasive powders during sample preparation.

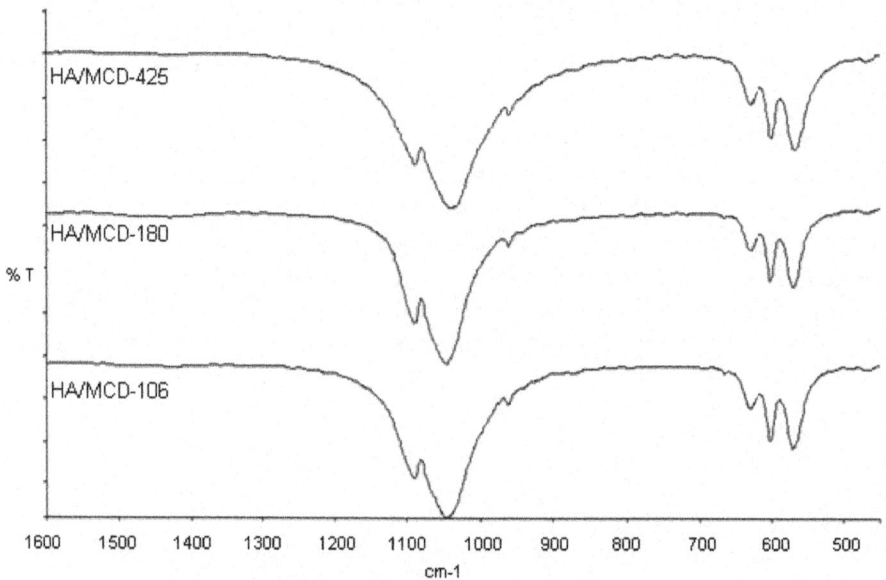

Figure 6. FTIR analysis of the CoBlast HA surface modifications.

CoBlast demonstrated the ability to deposit a well adhered HA coating with no major evidence of contamination with additional calcium phosphate phases. This is in contrast to the variety of crystalline phases and the presence of altered chemical functionality produced during the standard high temperature plasma spray deposition. The XRD and FTIR analysis does not suggest formation of any such impurities within the CoBlast samples indicating that the HA coating on the CoBlast samples has retained the same properties as the starting crystalline HA powder employed. Therefore it is anticipated that HA modified surfaces prepared using CoBlast should not exhibit the problems associated with the presence of impurities observed for HA modified surfaces prepared using standard high temperature plasma spray which are outlined above.

2.3. Cell Culture Analysis

2.3.1. Cell Proliferation

Osteoblasts are the key cells that are involved in the osteoconduction process. The success of the implantation is strongly influenced by how well the first phase of the attachment and adhesion of these cells will occur, which will then lead to the subsequent proliferation and differentiation upon contact with the implant surface. CoBlast surfaces have exhibited excellent osteoblast attachment and proliferation in vitro compared to their respective controls [21-23]. Biocompatability of the modified surfaces was determined via a (3-(4,5-dimethylthiazol-2-yl)-2,5-diphenyltetrazolium bromide) (MTT) assay and the osteoblast proliferation results are presented in Figure 7.

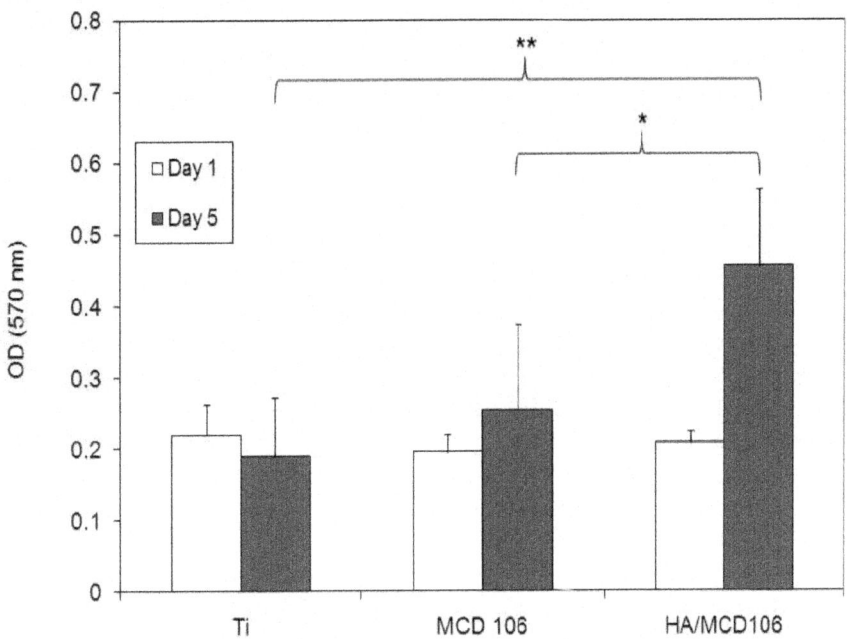

Figure 7. (3-(4,5-dimethylthiazol-2-yl)-2,5-diphenyltetrazolium bromide) (MTT) assay data for MG-63 cells on modified Ti surfaces over five days in vitro (*denotes $p < 0.05$ and **ascribes $p < 0.01$ determined using student's t-test).

Microblast and CoBlast surfaces prepared using the MCD 106 abrasive were selected for comparison. There was no significant difference in the cell proliferation between samples analyzed at day 1. However, at day 5 there was a significant increase ($p < 0.01$) in cell proliferation on the CoBlast coated substrate compared to the untreated titanium. A significant difference ($p < 0.05$) was also detected between the CoBlast sample and the microblast sample at day 5. The introduction of the bioactive HA layer through the CoBlast process was shown to further increase surface roughness and this combination of surface topography and bioactive surface chemistry was found to offer notably higher levels of cell proliferation after 5 days on the CoBlast surface. No evidence of cytotoxicity was observed using MG63 cells on any of the samples evaluated. However, the surface coating thickness and surface roughness of the CoBlast coatings was found to be lower compared to plasma HA coating in literature (20–300 μm thickness, 3–6 μm Ra)[34] with increased cell proliferation observed on plasma HA compared to grit blasted surfaces [1].

2.3.2. Cell Morphology

Surface modification techniques are extremely important for evoking desired cellular responses through tailoring the implants surface properties. The osteointegration process is greatly enhanced by these modifications, which in turn increases the long term success of the implant. As previously mentioned, HA is well known for its osteoconductive properties and its ability to influence cell adhesion and interaction at the implant-host interface. It is generally accepted that alongside chemical compatibility, the three dimensional surface topography (shape, size and surface texture) of the implant also influences early tissue response.

Figure 8 shows various images of MG63 cells that were cultured on the untreated titanium substrates, Figure 8a; microblasted MCD 106 surface, Figure 8b and CoBlast HA/MCD-106, Figure 8c after 24 hours.

Cells attached well to the untreated titanium surfaces and displayed a fibroblastic morphology synonymous with that of this osteoblastic cell line. The cells are typically polarized in one direction with the average cell length measuring roughly 60–80 μm (Figure 8a). Furthermore, lamellopodia and filopodia extensions (cytoskeletal organisation) from the main body of cells onto the surface are observed. The presence of these processes suggest good cell-substrate interactions where the cell is actively probing for specific topographical features and connects the cell to the substrate (via filopodia) from the lamellopida, which is the protrusion of their leading edge indicative of cell spreading and migration. However, the presence of numerous spherical cells indicates that not all cells are actively involved in spreading and migration. The cells cultured on MCD blasted surface (Figure 8b) display morphologies similar to that observed on the untreated titanium surfaces. The cells appeared to follow the contours of the MCD blasted surface where they would sit in the defects on the surface and align themselves along grooves or dents in a process referred to as 'contact guidance' [31]. The contact guidance effect has commonly been associated with increased cell proliferation and differentiation [31]. The surface roughness was determined to be 0.4, 0.5 and 0.7 μm for the blank Ti, microblast MCD-106 and the CoBlast HA/MCD-106 respectively and this obviously influenced the proliferation results observed here as mentioned earlier.

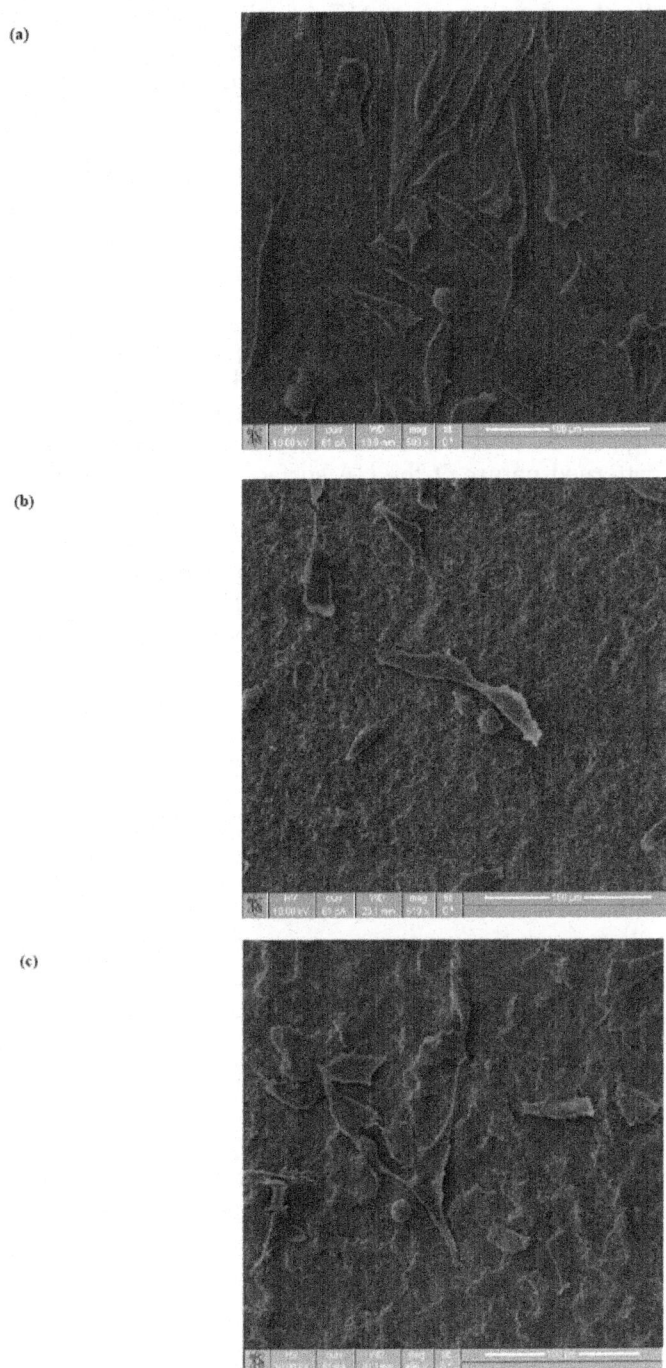

Figure 8. MG-63 osteoblasts cultured on (a) untreated Ti and (b) m1icroblasted MCD 106 treatment × 500 magnification and (c) HA/MCD-106 surfaces (CoBlast treatment) (×500 magnification).

Figure 8c shows an image of MG-63 cells cultured on the CoBlast after 24 hours, it can be seen that these differed from those of the untreated and microblasted titanium in terms of morphology where they had a polygonal shape rather than a polarized fibroblastic morphology indicating increased cell spreading. The cells also tended to align to the surface features created by the addition of the bioactive layer, which was earlier observed to increase the surface roughness (Figure 4). An abundance of lamellopodia and filopodia were present on the CoBlast surface. The addition of a HA coat to the Ti surface, resulted in a higher order of cell spreading and cell focal adhesion attachment compared to the microblasted surface (Figure 8b).

Research has proven that the manufacturing process and patterned topography have a significant influence on long term adherence and cell proliferation in vitro, irrespective of composition and surface roughness [32] and early controlled osteoblast alignment was demonstrated on patterned substrates [33]. Therefore, due to the patterned nature of the HA surface after CoBlast treatment, a more favorable surface resulted for cell spreading which subsequently lead to increased cell viability compared to the microblast and blank Ti surfaces. In vivo evaluation of CoBlast substrates prepared using an alumina abrasive have already demonstrated early stage lamellar bone growth [21]. This study demonstrates that by employing differing grades of MCD abrasive in the CoBlast process, greater control over surface topography can be achieved, which offers the capability to improve bone-implant contact in vivo. Further in vitro evaluations such as dissolution studies and bioactive studies in stimulated body fluid (SBF) must be investigated to support these capabilities.

3. EXPERIMENTAL SECTION

3.1. Materials

Titanium (Grade 5, Ti-6Al-4V) coupons (15 mm × 15 mm × 1 mm), were obtained from Lisnabrin Engineering Ireland. Hydroxyapatite [$Ca_{10}(PO_4)_6(OH)_2$] powder was sourced from S.A.I., France. The MCD apatite abrasives were all purchased from Himed Inc. (NY, USA). HPLC grade 1M hydrochloric acid (HCl), de-ionised water, isopropanol (IPA), ethanol, phosphate buffer solution (PBS), potassium bromide (KBr) FT-ir grade, Trypsin EDTA, MTT assay kit, ACS reagent grade dimethyl sulphoxide, paraformaldehyde, glutaraldehyde, osmium tetroxide and hexamethyldisilizane were all purchased from Sigma-Aldrich UK. MG-63 osteoblast cells were obtained from American Type Culture Collection, Rockville, MD, USA. Minimum Essential Medium (MEM), foetal calf serum, penicillin G sodium, streptomycin, amphotericin B were purchased from PAA Laboratories GmbH, Austria.

3.2. Sample Preparation

Prior to surface modification, the coupons were cleaned ultrasonically in 1M HCl and then in isopropanol to remove any contaminants. The CoBlast technique was used to modify the titanium, through deposition with a HA layer. The processing utilized twin microblast nozzles for the dopant/abrasive system. The HA (dopant) was deposited onto the Ti coupons using compressed air at a pressure of 90 psi, speed of 13 mm/sec and a working distance of approximately 20 mm. The MCD abrasives (either MCD-106, MCD-180 or MCD-425) were blasted out of the second nozzle at a pressure of 75 psi and at a working distance of 8 mm from the surface. The microblast surfaces where prepared under the same corresponding conditions as above, however no HA was delivered through the dopant nozzle. After the surface treatment step, each sample was ultrasonically washed in deionized water for 5 mins to remove any loose powder from the surface.

3.3. Surface Characterisation

The elemental composition of the powders and the coatings was carried out using a Jeol JSM 5510 SEM in conjunction with an INCA X-sight EDX spectroscopy detector (Oxford Instruments, Buckinghamshire, UK). Images were also taken using the same SEM system. Gravimetric analysis of the surface treatments on Ti was used to determine the coating mass using an Ohaus DV314C analytical balance by measuring the sample before and after an acid wash (ultrasonic treatment in 20 mL 1 M HCl for 10 mins). PXRD data was collected on a Siemens GAXRD diffractometer using $CuK\alpha 1$ radiation, with an anode current of 30 mA and an accelerating voltage of 40 keV. Data was collected in the range of 20 to 60 2θ degrees with a step size of 0.02 and a scan rate of 1 s per step. Coating thickness was measured using a PosiTector 6000 N thickness gauge (DeFelsko, NY, USA). An average of six readings was used to determine each value. The surface roughness (Ra) was determined using a Talsurf 10 surface profilometer (Talyor Hobson, UK). A Perkin Elmer Spectrum One FTIR was used to determine the structural fingerprint of the powders and the coatings. The coating was scrapped off, gently ground and pressed into a KBR disc (2% wt sample in KBR). Powders for FTIR analysis were prepared in a similar manner. FTIR spectra were recorded in the 1600–400-cm^{-1} range, with 4 cm^{-1} resolution using 20 scans and background subtraction. The spectra gave approximately 70–90% transmittance however the results are presented in an overlay fashion.

3.4. In Vitro Cell Culture

Sample modifications including CoBlast HA/MCD-106 and microblast MCD-106 and the blank Ti were evaluated for osteoconductivity and cytotoxicity using cell culture tests. Prior to cell culture analysis, each sample set was steam autoclaved at 121 °C for 20 minutes. MG-63 cells were used to assess cell proliferation. Cells were cultured in the MEM media supplemented with 10% foetal calf serum and antibiotic/antimycotic (penicillin G sodium 100 U/mL, streptomycin 100 μg/mL, amphotericin B 0.25 μg/mL) in 75 cm^3 tissue culture flasks. Cells were maintained in a humidified atmosphere with 5% CO_2 at 37 °C and were sub-cultured when they reached confluence using 0.25% Trypsin EDTA solution to provide adequate numbers of cells for the various in vitro culture studies undertaken.

3.4.1. Cell Proliferation

MG-63 cell attachment to the various treated and untreated Ti substrates was determined after 4 hours in culture using a commercial MTT assay and employing a modified Mosmann method [24]. Cells were seeded onto the samples at a concentration of $1x10^5$ cells/cm^2 and allowed to adhere during incubation at 37 °C in 5% CO_2 for 4 hrs. The MTT assay reagent was prepared as a 5 mg/mL stock solution in PBS, sterilized by filtration, and stored in the dark. An aliquot of the MTT stock solution (10% of total volume) was added to each well of a six well plate containing the samples (n = 4 for each sample type). After 3 hrs incubation at 37 °C in 5% CO_2, 200 μl of dimethyl sulfoxide was added to dissolve the formazan crystals. The solution was agitated for 15 min on a shaker to ensure adequate dissolution. The optical density of the formazan solutions was read by spectrophotometry using an ELISA plate reader (Tecan Sunrise, Tecan Austria) at 570 nm with the background absorbance value measured at 650 nm. The absorbance values recorded were determined to be proportional to the number of cells attached to the surface in each case. All data reported are expressed as mean ± standard deviation.

3.4.2. Cell Morphology

MG-63 cells were seeded onto each of the above substrates at a cell density of 5×10^5 cells/cm^2 in 6-well plates and were incubated for 24 hours. After cell culture, the samples were gently rinsed with PBS to remove any unattached cells and fixed in a modified Karnovsky's Fixative (2% paraformaldehyde/ 2% glutaraldehyde in PBS) for 1 hour. The samples were then rinsed in PBS and post-fixed in 1% osmium tetroxide and rinsed three times with PBS. The specimens were dehydrated by rinsing in an alcohol series (20, 30, 50, 70, 80, 90 and 95% ethanol), and finally rinsing 3 times in 100% ethanol. The samples were then were chemically dried in hexamethyldisilizane (HMDS) overnight. A 50 nm layer of gold-palladium was deposited onto the substrates using a Polaron E5000 SEM Sputter Coating Unit. The sputtering conditions used a set voltage of 1.4 kV, with a plasma current of 18 mA (argon gas), a deposition time of 2 minutes at a vacuum pressure of 0.05 Torr. The samples were then analyzed using the Jeol JSM 5510 SEM and subsequently using a FEI Quanta 200 Focused Ion Beam and SEM in backscatter electron mode.

4. CONCLUSIONS

Detailed surface studies have shown that the combination of an apatite abrasive and a HA dopant in the CoBlast process produces surfaces with a combination of optimized apatite chemistry and controlled surface structure. The CoBlast process has the ability to retain the chemistry of the starting HA material. This offers advantages over conventional high temperature plasma processing which alters the HA material from its desired chemical, structural and dissolution requirements for its use as an in vivo implant material. The study also shows that employing MCD abrasives offer an alternative to alumina for deposition using CoBlast process. In vitro studies clearly show that increased roughness of treated surfaces favors enhanced cell proliferation and the CoBlast process offers the ability to tailor the surface texture to produce an optimized surface for osteointegration of a HA modified implant. Enhanced cell proliferation was observed for CoBlast modified surfaces compared to the microblasted surface. The ability of the CoBlast technology to offer diversity in modifying surface topography is clearly shown in this and previous studies and represents foundation work, which supported by bioactivity studies and in vivo trials, offers exciting new prospects in tailoring the properties of medical devices for applications ranging from dental to orthopedic settings.

Acknowledgments

The authors would like to acknowledge EnBio for supplying the CoBlast samples for this study.

REFERENCES AND NOTES

1. Borsari, V.; Giavaresi, G.; Fini, M.; Torricelli, P.; Salito, A.; Chiesa, R.; Chiusoli, L.; Volpert, A.; Rimondini, L.; Giardino, R. Physical characterization of different-roughness titanium surfaces, with and without hydroxyapatite coating and their effect on human osteoblast-like cells. J. Biomed. Mater. Res. Part B 2005, 75B, 359-368.

2. Stoch, A.; Jastrze, B.W.; Dlugon, E.; Lejda, W.; Trybalska, B.; Stoch, G.J.; Adamczyk, A. Sol-gel derived hydroxyapatite coatings on titanium and its alloy Ti6Al4V. J. Mol. Struct. 2005, 744, 633-640.

3. Oh, I.H.; Nomura, N.; Chiba, A. Microstructures and bond strengths of plasma-sprayed hydroxyapatite coatings on porous titanium substrates. J. Mater. Sci. Mater. Med. 2005, 16, 635-640.

4. Lu, Y.P.; Li, M.S.; Li, S.T.; Wang, Z.G.; Zhu, R.F. Plasma-sprayed hydroxyapatite + titania composite bond coat for hydroxyapatite coating on titanium substrate. Biomaterials 2004, 25, 4393-4403.

5. Wennerberg, A.; Ektessabi, A.; Albrektsson, T.; Johansson, C.; Andersson, B.A. 1-year follow-up of implants of differing surface roughness placed in rabbit bone. Inter. J. Oral Maxillofac. Implants 1997, 12, 486-494.

6. Abron, A.; Hopfensperger, M.; Thompson, J.; Cooper, L.F. Evaluation of a predictive model for implant surface topography effects on early osseointegration in the rat tibia model. J. Prosthet. Dent. 2001, 85, 40-46.

7. Nakada, H.; Sakae, T.; Legeros, R.Z.; Legeros, J.P.; Suwa, T.; Numata, Y.; Kobayashi, K. Early tissue response to modified implant surfaces using back scattered imaging. Implant Dent. 2007, 16, 281-289

8. Gil, F.J.; Planell, J.A.; Padros, A.; Aparicio, C. The effect of shot blasting and heat treatment on the fatigue behavior of titanium for dental implant applications. Dent. Mater. 2007, 23, 486-491.

9. Chen, J.; Wolke, J.G.C.; De Groot, K. Microstructure and crystallinity in hydroxyapatite coatings. Biomaterials 1994, 15, 396-399.

10. Gross, K.A.; Berndt, C.C.; Herman, H. Amorphous phase formation in plasma-sprayed hydroxyapatite coatings. J. Biomed. Mater. Res. 1998, 39, 407-414.

11. Gross, K.A.; Berndt, C.C. Thermal processing of hydroxyapatite for coating production. J. Biomed. Mater. Res. 1998, 39, 580-587.

12. Heimann, R.B.; Wirth, R. Formation and transformation of amorphous calcium phosphates on titanium alloy surfaces during atmospheric plasma spraying and their subsequent in vitro performance. Biomaterials 2006, 27, 823-831.

13. Weng, J.; Liu, Q.; Wolke, J.G.; Zhang, X.; De Groot, K. Formation and characteristics of the apatite layer on plasma-sprayed hydroxyapatite coatings in simulated body fluid. Biomaterials 1997, 18, 1027-1035.

14. Li, H.; Li, Z.X.; Li, H.; Wu, Y.Z.; Wei, Q. Characterization of plasma sprayed hydroxyapatite/ ZrO_2 graded coating. Mater. Design 2009, 30, 3920-3924.

15. Katto, M.; Kurosawa, K.; Yokotani, A.; Kubodera, S.; Kameyama, A.; Higashiguchi, T.; Nakayama, T.; Tsukamoto, M. Poly-crystallized hydroxyapatite coating deposited by pulsed laser deposition method at room temperature. Appl. Surf. Sci. 2005, 248,365-368.

16. Hong, Z.; Luan, L. Paik, S.E.; Deng, B.; Ellis, D.E.; Ketterson, J.B. Mello, A.; Eon, J.G., Terra, J., Rossi, A. Crystalline hydroxyapatite thin films produced at room temperature — An opposing radio frequency magnetron sputtering approach. Thin Solid Films 2007, 515, 6773-6780.

17. Stoch, A.; Brozek, A.; Kmita, G.; Stoch, J.; Jastrzebski, W.; Rakowska, A. Electrophoretic coating of hydroxyapatite on titanium implants. J. Mole. Struct. 2001, 596, 191-200.

18. Mano, T.; Ueyama, Y.; Ishikawa, K.; Matsumura, T.; Suzuki, K. Initial tissue response to a titanium implant coated with apatite at room temperature using a blast coating method. Biomaterials 2002, 23, 1931-1926.

19. Gbureck, U.; Masten, A.; Probst, J.; Thull, R. Tribochemical structuring and coating of implant metal surfaces with titanium oxide and hydroxyapatite layers. Mater. Sci. Eng. C 2003, 23, 461-465.

20. Ishikawa, K.; Miyamoto, Y.; Nagayama, M.; Asaoka, K. Blast coating method: New method of coating titanium surface with hydroxyapatite at room temperature. J. Biomed. Mater. Res. 1997, 38,129-134.

21. O'Hare, P.; Meenan, B.J.B.; George, A.; Byrne, G.; Dowling, D.; Hunt, J.A. In vitro and in vivo response of hydroxyapatite surfaces deposited via a novel co-incident microblasting technique for improved orthopaedic implant performance. Biomaterials 2010, 31, 515-522.

22. O'Neill, L.; O'Sullivan, C.; O'Hare, P.; Sexton, L.; Keady, F.; O'Donoghue, J. Deposition of substituted apatites onto titanium surfaces using a novel blasting process. Surf. Coat. Technol. 2009, 204, 484-488.

23. O'Sullivan, C.; O'Hare, P.; O'Leary, N.D.; Crean, A.M.; Ryan, K.; Dobson, A.D.; O'Neill, L.D. Deposition of substituted apatites with anticolonizing properties onto titanium surfaces using a novel blasting process. J. Biomed. Mater. Res. Part B 2010, 95,141-149.

24. Mosmann, T. Rapid colorimetric assay for cellular growth and survival: Application to proliferation and cytotoxicity assays. J. Immunol. Methods 1983, 65, 55-63.

25. Kim, T.N.; Feng, Q.L.; Kim, J.O.; Wu, J.; Wang, H.; Chen, G.C.; Cui, F.Z. Antimicrobial effects of metal ions (Ag^+, Cu^{2+}, Zn^{2+}) in hydroxyapatite. J. Mater. Sci.: Mater. Med. 1998, 9, 129-134.

26. Fathi, M.H.; Hanifi, A.; Mortazavi, V. Preparation and bioactivity evaluation of bone-like hydroxyapatite nanopowder. J. Mater. Proc. Technol. 2008, 202,536-542.

27. Varma, H.K.; Babu, S.S. Synthesis of calcium phosphate bioceramics by citrate gel pyrolysis method. Ceram. Int.2005, 31, 109-114.

28. Pleshko, N.; Boskey, A.; Mendelsohn, R. Novel infrared spectroscopic method for the determination of crystallinity of hydroxyapatite minerals. Biophy. J.1991, 60,786-793.

29. Legeros, R.Z.; Daculsi, G.; Orly, I.; Gregoire, M. Substrate surface dissolution and interfacial biological minerals. In The Bone-Biomaterial Interface; Davies, J.E., Ed.; University of Toronto: Toronto, Canada, 1991; pp. 76-88.

30. Masmoudi, M.; Assoul, M.; Wery, M.; Abdelhedi, R.; El Halouani, F.; Monteil, G. Friction and wear behaviour of cp Ti and Ti6Al4V following nitric acid passivation. Appl. Surf. Sci. 2006, 253, 237-2243.

31. Anselme, K. Osteoblast adhesion on biomaterials. Biomaterials 2000, 21, 667-681.

32. Bigerelle, M.; Anselme K. Statistical correlation between call adhesion and proliferation on biocompatible metallic materials. J. Biomed. Mater. Res. A 2005, 75, 530-540.

33. Puckett, A.; Pareta, R.; Webster, T.J. Nano rough micron patterned titanium for directing osteoblast morphology and adhesion. Inter. J. Nanomed. 2008, 2, 229-241.

34. Sun, L.; Berndt, C.C.; Gross, K.A.; Kucuk, A. Material fundamentals and clinical performance of plasma-sprayed hydroxyapatite coatings: A review. J Biomed. Mate. Res. 2001, 58, 570-592.

This page left intentionally blank.

INDEX

This page left intentionally blank.